PROPERTIES OF ENGINEERING MATERIALS

PROPERTIES OF

ENGINEERING MATERIALS

Theory, worked examples and problems

Gladius Lewis
B.Sc.(Eng.), M.Sc., Ph.D., M.Inst.B.E., A.M.I.Corr.T.

Lecturer in the Faculty of Engineering,
University of Zimbabwe, Salisbury

First published 1981 by
THE MACMILLAN PRESS LTD
London and Basingstoke
Companies and representatives
throughout the world

Printed in Hong Kong

ISBN 0 333 30741 0

CONTENTS

PREFACE

A knowledge of the properties of engineering materials is crucial to
the safe design and performance of components in service. The
subject has, however, not often been treated in a manner which lends
itself to easy solutions of problems by undergraduate students. The
purpose of this book is to illustrate, in a concise form, the basis
of the properties of engineering materials and hence to aid problem-
solution. For each topic covered, the format consists of a present-
ation of the relevant theory in a compact fashion, omitting the wider
treatment, followed by a number of worked examples and problems for
the reader to solve on his own.

The main reason for this approach is that there are well-written
textbooks available which the student can consult for both back-
ground material and detailed exposition. This book therefore assumes
some basic knowledge of the subject and should be used in conjunction
with such textbooks as are listed in appendix I.

I wish to express my thanks to the Senates of the Universities of
London, Cambridge and Nottingham for permission to reproduce
questions from their past examination papers. The range of questions
presented here will make the book useful to all levels of technical
college, polytechnic and undergraduate university students. I alone
am responsible for the solutions and for the unacknowledged questions.

GLADIUS LEWIS

1   MECHANICAL PROPERTIES

1.1 TRUE STRESS AND TRUE STRAIN

The nominal (or engineering) stress-strain curve does not give a true
indication of the deformation characteristics of a material because

(1)   it is based entirely on the original dimensions of the tensile
       testpiece, and
(2)   ductile material that is pulled in tension becomes unstable and
       necks down during the course of the tensile test.

In order to take into account the change in the original
dimensions of the specimen during the tensile test, true stress-true
strain curves are plotted as shown in figure 1.1.

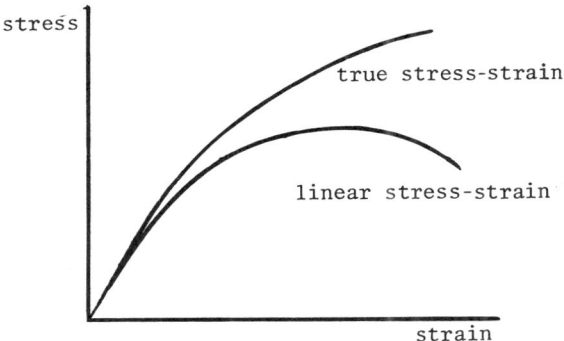

Figure 1.1

Nominal (or engineering) stress, $S = P/A_0$

Nominal (or engineering) strain, $e = (L - L_0)/L_0$

True stress, $\sigma = P/A_i$

True strain, $\varepsilon = \displaystyle\int_{L_0}^{L} \frac{dL}{L} = \ln\left(\frac{L}{L_0}\right)$

where P = applied load, A_0 = original cross-sectional area, A_i =
instantaneous cross-sectional area, L_0 = initial gauge length, L =
instantaneous gauge length.

1

Now $1 + e = L/L_0$, hence $\varepsilon = \ln(1 + e)$ and

$$\text{true stress, } \sigma = \frac{P}{A_i} = \frac{P}{A_0} \times \frac{A_0}{A_i}$$

Because plastic deformation occurs by a process of shear there is essentially no volume change in the specimen during deformation; hence $A_i L = A_0 L_0$ so we can write

$$\text{true stress } \sigma = \frac{P}{A_0} \times \frac{L}{L_0} = \frac{P}{A_0}(1 + e)$$

or $\quad P = \dfrac{A_0 \sigma}{1 + e}$

For a maximum, $dP/de = 0$, that is

$$\frac{dP}{de} = \frac{(1 + e)A_0 \frac{d\sigma}{de} - A_0 \sigma}{(1 + e)^2} = 0$$

$$\frac{d\sigma}{de} = \frac{\sigma}{1 + e} \qquad\qquad\qquad (1.1)$$

For most metals, $\sigma = ae^b$, where a and b are material constants; so

$$\frac{d\sigma}{de} = abe^{b-1} \qquad\qquad\qquad (1.2)$$

Equating equations 1.1 and 1.2 and putting $\sigma = ae^b$, we have

$$abe^{b-1} = \frac{ae^b}{1 + e}$$

whence

$$e = \frac{b}{1 - b}$$

From Considère's construction (figure 1.2), using similar triangles ABC and ADE, we obtain maximum tensile stress, S_{max}, from

$$\frac{S_{max}}{1} = \frac{\sigma}{1 + e}$$

i.e. $S_{max} = \dfrac{\sigma}{1 + e} = \dfrac{a[b/(1 - b)]^b}{1 + [b/(1 - b)]} = ab^b(1 - b)^{1-b}$

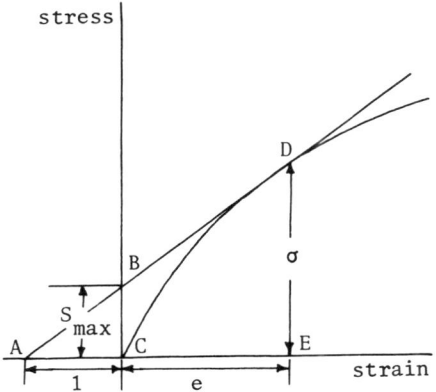

Figure 1.2

1.2 HARDNESS

Hardness can be defined as resistance to localised indentation of a surface by a standard indenter under standard loading conditions. The smaller the indentation the harder the material.

The principle of indentation hardness measurement is simple. A suitably shaped indenter is driven into the surface of the testpiece under a constant load. As it penetrates, the contact area between the testpiece and indenter increases and hence the contact stress decreases. When this stress has decreased to the yield strength of the material, the penetration stops. Hardness is then expressed as the ratio of the indenting force to some measure of the indentation size generated. Three methods are in common use, differing only in the shape and size of the indenter.

The Vickers hardness (HV) test uses a square-based diamond pyramid, having an included angle between opposite faces of 136°. The loads used vary between 1 kgf and 50 kgf for about 15 s. Then

$$HV = \frac{2P \sin (\theta/2)}{d^2} = \frac{1.854P}{d^2} \text{ kgf mm}^{-2}$$

where P = load, θ = 136°, d = average length of the diagonal of the square generated by the intersection of the indenter and the material surface.

The Brinell hardness (HB) test uses a hardened steel ball, 5 mm or 10 mm in diameter, as the indenter. The load is usually either 500 kgf or 3000 kgf, applied for about 10 s. Then

3

$$HB = \frac{P}{(\pi D/2)[D - \sqrt{(D^2 - d^2)}]} \ kgf \ mm^{-2}$$

where P = load, D = steel ball diameter, d = indentation diameter.

In the Rockwell hardness (HR) test a small indentation under a small load (usually 10 kgf) is made and then the load is increased. The hardness is then the ratio of the indentation under the test load to that under the pre-load. Three scales of HR are used, depending on the type of indenter and test load, as follows.

Scale	Indenter Type	Applied Load (kgf)
HRA	Diamond	60
HRB	1.6 mm diameter steel ball	100
HRC	Conical diamond	150

1.3 MEYER HARDNESS ANALYSIS

Meyer showed that the following relationship exists between the applied load, P, and indentation diameter, d, in a Brinell hardness test

$$P = ad^n$$

where the constant, a, represents the resistance to first penetration and is a function of the ball size used, and the constant, n, is a function of the ability of the material to work harden (2.0<n<2.5).

Example 1.1

The following results were obtained during the first part of a tensile test on an aluminium specimen, gauge length 69 mm and diameter 13.82 mm.

Load (kN)	0	5.0	7.5	10.0	12.5	15.0
Extension (mm)	0	0.0329	0.0493	0.0658	0.0960	0.1852

Plot a load-extension diagram and find the values of (a) the Young's modulus of elasticity and (b) the 0.1% proof stress for this material.

The load-extension graph is given in figure 1.3.

(a) Original cross-sectional area = $\frac{\pi}{4}(13.82)^2$ = 150 mm^2

In the linear portion of this graph, extension is 0.0658 mm per 10 kN load. Thus

$$increase \ in \ stress = \frac{10}{150} \ \frac{kN}{mm^2} = 66.7 \ MN \ m^{-2}$$

$$increase \ in \ strain = \frac{0.0658}{69} = 0.000954$$

4

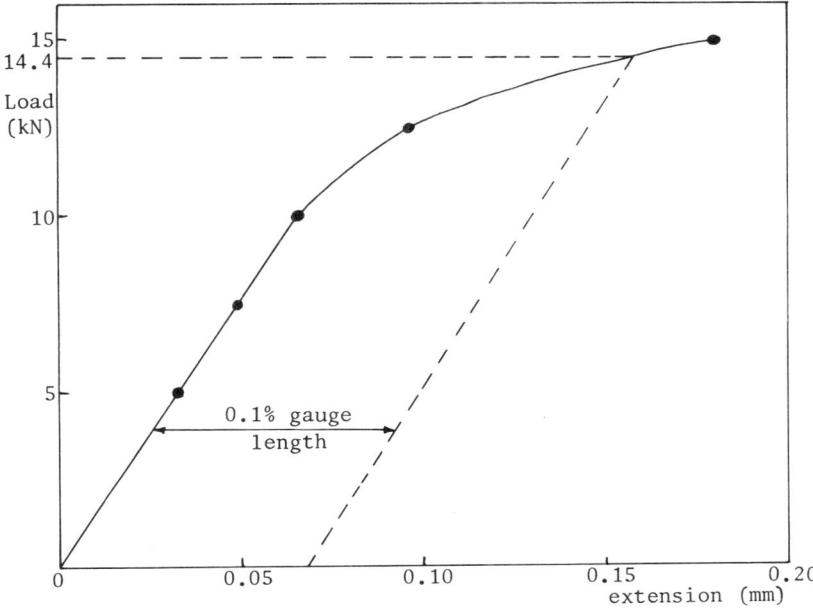

Fig. 1.3

therefore

$$\text{Young's modulus of elasticity, } E = \frac{66.7 \times 10^6}{0.000954} \text{N m}^{-2} = 70 \text{ GN m}^{-2}$$

   (b) 0.1% gauge length is 0.069 mm; corresponding load, from figure 1.1, is 14.4 kN; thus

$$0.1\% \text{ proof stress} = \frac{14.4}{150} \frac{\text{kN}}{\text{mm}^2} = 96 \text{ MN m}^{-2}$$

Example 1.2

A frictionless uniaxial compression test on an annealed aluminium cylinder initially 17 mm diameter and 25 mm long gave the following results.

Load (kN)	Reduction in Height (mm)
11.0	0.1
16.5	0.7
22.0	1.9
27.5	3.6
33.0	5.1
38.5	6.6
44.0	8.0

Determine the tensile strength and the value of the linear strain at which necking would start in a tensile test on the material.

(Cambridge)

Cross-sectional area of cylinder = 227 mm^2. We now calculate the true stress, $\sigma = (P/A)(1 + e)$, and linear strain, e, for each result.

True Stress (MN m^{-2})	Linear Strain
48.69	0.004
74.74	0.028
104.26	0.076
138.54	0.144
175.06	0.204
214.37	0.264
255.81	0.320

The relevant relationship is $\sigma = ae^b$ and the values of the constants a and b are obtained from a plot of log σ against log e (figure 1.4). At necking, linear strain = b and the tensile strength is $ab^b(1 - b)^{1-b}$.

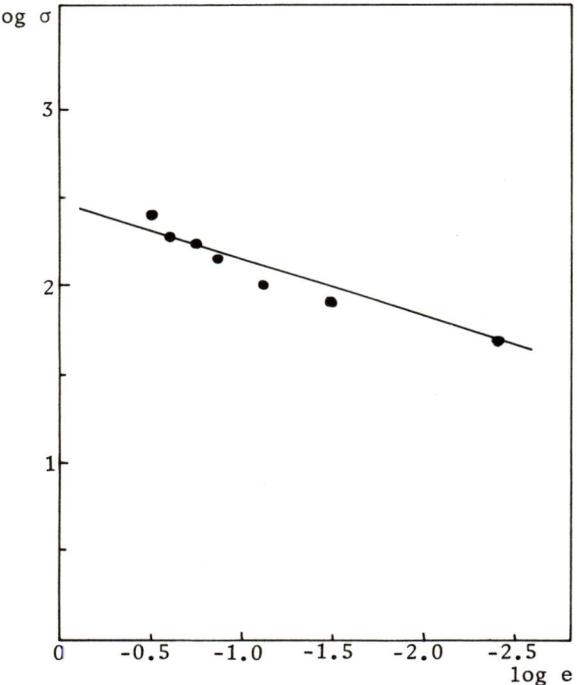

Fig. 1.4

In this case, we have (1) linear strain at necking = slope of figure 1.4 = 0.33. (2) intercept on figure 1.4 = 302 hence

6

tensile strength = $302 \times 0.33^{0.33} \times 0.67^{0.67}$ = 160 MN m^{-2}

Example 1.3

A specimen of mild steel of diameter 14.3 mm and gauge length 51 mm is tested in tension and the elongation at fracture is 35.2%. A second specimen of the same material, diameter 5.4 mm and gauge length 60 mm, fractures at an elongation of 24.2%. Determine the percentage elongation of a specimen of the same material of rectangular section 25.4 mm by 12.7 mm and of 100 mm gauge length.

Applying Barba's law to specimens 1 and 2, that is

$$\% \text{ elongation} = 100\left(\frac{c\sqrt{A}}{L} + b\right)$$

where A = cross-sectional area, L = gauge length, c and b are material constants, we have

$$35.2 = 100\left[\frac{c\sqrt{(\pi \times 14.3^2/4)}}{51} + b\right]$$

and $$24.2 = 100\left[\frac{c\sqrt{(\pi \times 5.4^2/4)}}{60} + b\right]$$

whence b = 0.19 and c = 0.65.

$$\text{Required percentage elongation} = 100\left[\frac{0.65\sqrt{(25.4 \times 12.7)}}{100} + 0.19\right]$$

$$= 30.67\%$$

Example 1.4

Two testpieces were cut from a piece of mild steel plate and, when tested in simple tension, gave the following results.

Specimen	Cross-section	Gauge Length	% Elongation
1	50 mm × 6 mm	150 mm	26.2
2	75 mm × 6 mm	100 mm	33.2

What is the probable percentage elongation for a piece of the same plate 60 mm × 6 mm when tested on a gauge length of 200 mm?

Barba's law states that

$$\% \text{ elongation} = 100\left(\frac{c\sqrt{A}}{L} + b\right)$$

In this example, we have

$$26.2 = 100\left(\frac{c\sqrt{300}}{150} + b\right)$$

and $$33.2 = 100\left(\frac{c\sqrt{450}}{100} + b\right)$$

7

whence b = 0.178 and c = 0.730.

For the case where the cross-sectional area is 360 mm^2 and the gauge length is 200 mm

$$\% \text{ elongation} = 100 \left(\frac{0.730\sqrt{360}}{200} + 0.178 \right) = 24.73\%$$

Example 1.5

A specimen of gauge length L_0 is loaded in tension until its length is $2L_0$. A second specimen of gauge length L_0 is compressed until its length is $L_0/2$. Show that the values of the true strain for the two specimens are equal but the engineering strains are not.

$$\text{True strain in tension, } \varepsilon_t = \ln \left(\frac{2L_0}{L_0} \right) = \ln 2 = 0.69 \tag{1}$$

$$\text{true strain in compression, } \varepsilon_c = \ln \left(\frac{L_0/2}{L_0} \right) = \ln 1/2 = -0.69 \tag{2}$$

$$\text{engineering strain in tension, } e_t = \frac{2L_0 - L_0}{L_0} = 1.0 \tag{3}$$

$$\text{engineering strain in compression, } e_c = \frac{(L_0/2) - L_0}{L_0} = -0.5 \tag{4}$$

Thus the true strains (equations 1 and 2) are numerically equal but the engineering strains (equations 3 and 4) are different.

Example 1.6

For an annealed metal the true stress-true strain relationship in a tensile test is approximated by

$$\sigma = K\varepsilon^n$$

where K and n are positive constants. (a) Use this expression to find the true strain at the maximum load. (b) In terms of K and n what is the work done per unit volume in straining the metal to maximum load in a tensile test?

(Cambridge)

(a) At necking, (i.e. at maximum load)

$$\varepsilon = n$$

(b) $$\frac{\text{work}}{\text{volume}} = \int_0^n \frac{\text{load} \times \text{extension}}{\text{area} \times \text{length}}$$

$$= \int_0^n \sigma \, d\varepsilon$$

$$= \int_0^n K\varepsilon^n \, d\varepsilon$$

$$= \frac{K\varepsilon^{n+1}}{n+1}$$

$$= \frac{Kn^{n+1}}{n+1} \, \text{J m}^{-3}$$

Example 1.7

The stress-strain curve for an alloy may be expressed as

$$\sigma = 430e^{0.5} \text{ MN m}^{-2}$$

where σ = true stress and e = linear strain. (a) What is the true strain at the start of necking? (b) Calculate the work per unit volume absorbed in fracturing the alloy if the strain at fracture is 0.3.

(a) At the start of necking, linear strain, e_n = 0.5, so

true strain, $\varepsilon_n = \ell n(1 + 0.5) = 0.41$

(b) Work $= \int_0^{e_f} \text{load} \times \text{extension}$

where e_f = fracture strain; then

$$\frac{\text{work}}{\text{volume}} = \int_0^{e_f} \frac{\text{load} \times \text{extension}}{\text{area} \times \text{length}} = \int_0^{e_f} \sigma \, de$$

$$= \int_0^{0.3} 430e^{0.5} \, de = 47 \text{ MJ m}^{-3}$$

Example 1.8

The relationship between true stress, σ, and strain, ε, for a sample of pure aluminium may be represented by

$$\sigma = 90\varepsilon^{0.3} \text{ MN m}^{-2}$$

For a specimen of this material undergoing a tensile test, what is (a) the linear strain at which necking commences? (b) the tensile strength? (c) the work per unit volume necessary to deform the specimen up to the onset of necking? (Cambridge)

9

(a) The linear, e, and true, ε, strains are related by

$$\varepsilon = \ell n(1 + e)$$

At necking, $\varepsilon = 0.3$ so that

$$\exp(0.3) = 1 + e$$

or   e = 0.35

(b) Tensile strength, $S_{max} = 90 \times 0.35^{0.35} \times 0.65^{0.65} = 47$ MN m^{-2}

(c) $\dfrac{\text{Work}}{\text{volume}} = \displaystyle\int_0^{0.3} \dfrac{\text{force} \times \text{extension}}{\text{area} \times \text{length}} = \int_0^{0.3} \sigma\, d\varepsilon = \int_0^{0.3} 90\varepsilon^{0.3}$

$$= 14.5 \text{ MJ m}^{-3}$$

Example 1.9

The true stress-nominal strain relationship for an annealed metal with modulus of elasticity of 7×10^4 MN m^{-2} can be approximated by

$$\sigma = 400e^{0.4} \text{ MN m}^{-2}$$

where σ = true stress, e = linear strain. Find the ratio of the plastic to the elastic strain at the maximum tensile load. (Cambridge)

Maximum tensile strength. $S_{max} = 400 \times 0.4^{0.4} \times 0.6^{0.6}$

$$= 204 \text{ MN m}^{-2}$$

Thus elastic strain, $e_e = \dfrac{204}{7 \times 10^4} = 29 \times 10^{-4}$

At the start of the plastic stage, the plastic strain, $e_p = 0.4$. Thus

$$\text{ratio } \frac{e_p}{e_e} = \frac{0.4}{29 \times 10^{-4}} = 138$$

Example 1.10

The following hardness data were obtained on an annealed copper specimen using a 10 mm ball diameter.

Load (kg)	Diameter of Indentation (mm)
500	4.4
1000	5.4
1500	6.2

Determine the Meyer constants.

Meyer's law states

$$P = kd^n \tag{1}$$

where P = load, d = diameter of indentation, k and n are material constants.

Equation 1 can be written

$$\log P = \log k + n \log d$$

Thus by plotting the hardness data as log P against log d, both k and ı can be obtained. The slope of this graph (figure 1.5) gives n = 3.3 and the intercept = 0.68 whence k = 4.8.

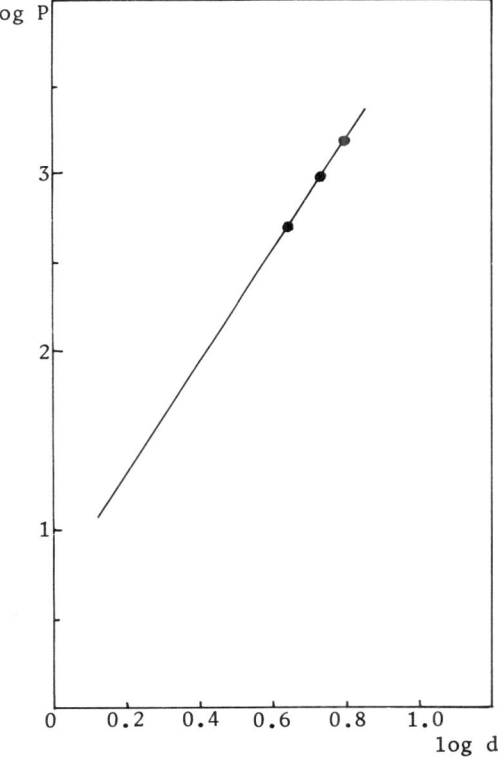

Fig. 1.5

Example 1.11

It has been shown that the relationship between Rockwell C hardness,

11

R_c, and Brinell hardness, HB, values may be expressed by an equation of the form

$$R_c = C_1 - \frac{C_2}{\sqrt{(HB)}}$$

where C_1 and C_2 are constants. Using the hardness-conversion data for for steel given below, evaluate the constants in this equation and comment on the goodness-of-fit.

Hardness (kgf mm^{-2})

Rockwell C	Brinell
58.5	600
42.7	397
27.1	265
11.0	195

By plotting R_c against $1/\sqrt{(HB)}$ (figure 1.6) we have C_1 = intercept of graph = 121.4 and C_2 = slope of graph = 1542. Thus the equation becomes

$$R_c = 121.4 \left(1 - \frac{12.7}{\sqrt{(HB)}}\right) \tag{1}$$

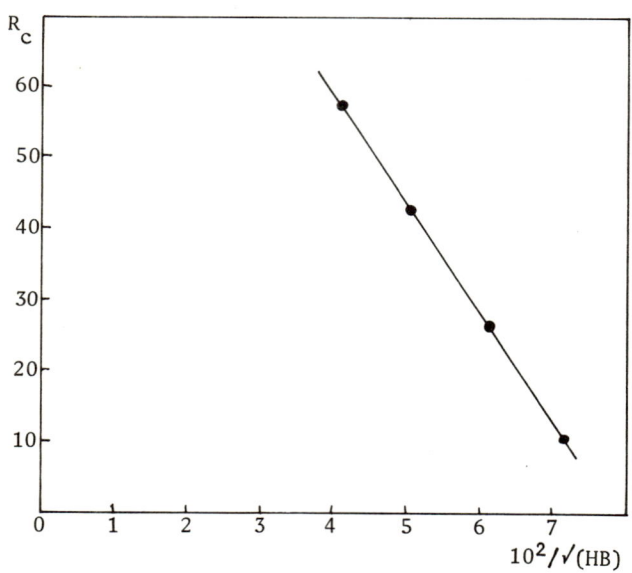

Fig. 1.6

For each value of HB given, R_c was calculated using the above equation and then the standard error of estimate calculated.

HB	Experimental R_c	Calculated R_c	Deviation^2
600	58.5	58.3	0.04
397	42.7	43.7	1.00
265	27.1	26.7	0.16
195	11.0	10.9	0.01

$$\text{Standard error of estimate} = \sqrt{\left(\frac{0.04 + 1.00 + 0.16 + 0.01}{4 - 2}\right)} = 0.78$$

The calculated coefficient of correlation of the fitted linear equation (equation 1) is 0.99. The values of the two indices, standard error of estimate and coefficient of correlation, indicate that the linear equation fits the given data adequately.

Example 1.12

A Vickers diamond hardness test is made on a steel with a load of 30 kgf and a square pyramidal indenter of included angle 136°. The impression measures 0.654 mm across the diagonals. Determine the Vickers hardness number.

What particular load should be used to investigate the hardness of a particular grain in this material if the grain diameter is 0.05 mm and its hardness is not expected to be less than that of the bulk specimen?
(Cambridge)

(a) The Vickers hardness number, HV, is given by the expression

$$HV = \frac{1.854P}{d^2}$$

for an included angle, θ, of 136°. In this case

$$HV = \frac{1.854 \times 30}{0.654^2} = 130 \text{ kgf mm}^{-2}$$

(b) For a hardness of 130 kgf mm^{-2} the load on the grain can be evaluated using the expression given above, that is

$$130 = \frac{1.854P}{0.05^2}$$

whence P = 0.18 kgf.

Example 1.13

In general the Brinell hardness of a given material is a constant for a given applied indenter load'. Examine this statement with respect to the following data from Brinell hardness tests.

13

Applied Load (kg)	Indenter Diameter (mm)	Indentation Diameter (mm)
3000	10	4.75
1470	7	3.33
750	5	2.35

The expression for Brinell hardness is

$$HB = \frac{P}{(\pi D/2)[D - \sqrt{(D^2 - d^2)}]}$$

The calculated values of HB for the given P, D and D are thus

Test No.	P (kgf)	D (mm)	d (mm)	HB (kgf mm^{-2})
1	3000	10	4.75	159.12
2	1470	7	3.33	158.60
3	750	5	2.35	162.75

The values in the last column are approximately constant (161 ± 2), so the statement in the question is confirmed.

Example 1.14

It has been shown that $HV = 2.9Y_r$, where Y_r = yield stress measured at a strain of 0.08 more than that at which the hardness is measured, and HV = diamond pyramid hardness of a metal in a particular state of strain. A certain metal has a stress-strain curve given by

$$\sigma = ae^{0.32}$$

where σ = true stress, e = linear strain, a is a constant. Determine, from first principles, an expression for the tensile strength of the metal in terms of the measured value of HV for the specimen. (Cambridge)

A and ℓ are the cross-sectional area and length, respectively. Plastic deformation causes negligible change in volume so that

$$A\ell = A_0\ell_0$$

or $$A = \frac{A_0\ell_0}{\ell} = \frac{A_0}{1 + e}$$

True stress, $\sigma = (P/A_0)(1 + e)$ or

$$P = \frac{A_0\sigma}{1 + e} \tag{1}$$

From which

$$\frac{dP}{de} = \frac{A_0}{(1 + e)^2}\left[(1 + e)\frac{d\sigma}{de} - \sigma\right] \tag{2}$$

14

The maximum load, P_{max} will occur at a strain for which $dP/de = 0$, i.e.

$$(1 + e) \frac{d\sigma}{de} = \sigma$$

$$\frac{d\sigma}{de} = \frac{\sigma}{1 + e} \qquad (3)$$

True stress-linear strain relationship for metals is $\sigma = ae^b$, then

$$\frac{d\sigma}{de} = abe^{b-1} \qquad (4)$$

From equations 3 and 4, $e = b/(1-b)$, then

maximum strength, $S_{max} = \dfrac{\sigma}{1 + e} = \dfrac{ab/(1 - b)^b}{1 + [b/(1 - b)]} = ab^b(1 - b)^{1-b}$

The diamond indenter produces an average strain of 0.08 so that the yield stress is $\sigma_y = a(0.08)^b$. Now $HV = 2.9\sigma_y$, therefore

$$HV = 2.9a(0.08)^b$$

or $\quad a = \dfrac{HV(12.5)^b}{2.9}$

Hence

$$S_{max} = \frac{HV}{2.9} (12.5b)^b(1 - b)^{1-b}$$

With $b = 0.32$, we now have

$$S_{max} = \frac{HV}{2.9} (12.5 \times 0.32)^{0.32}(0.68)^{0.68} = 0.413HV \text{ kgf mm}^{-2}$$

or $\quad S_{max} = 4.05HV \text{ MN m}^{-2}$

Example 1.15

Determine the value of the maximum contact stress at the initial instant when the steel ball of a Brinell testing machine is being pressed into a plane component under test. The diameter of the ball is 10 mm, the force is 10 N. The ball and component are made of steel, with modulus of elasticity of 200 GN m^{-2}.

The formula for the maximum stress for the given type of contact is given as

$$S'_{max} = 0.388 \sqrt[3]{\left(\frac{PE^2}{R^2}\right)}$$

where P = force, R = ball radius, E = modulus of elasticity. In this case, P = 10 N, R = 5 mm, E = 200 GN m^{-2}, so that

$$S'_{max} = 0.388 \sqrt[3]{\left(\frac{10(200 \times 10^9)^2}{(0.005)^2}\right)} = 980 \text{ MN m}^{-2}$$

PROBLEMS

(1) Explain the engineering significance in an indentation hardness test of the relationship between the load, P kgf, on the indenting ball and the diameter, D mm, of the ball, given by P/D^2 = constant.

(2) What do the constants in the Meyer relationship for the indentation hardness signify?

   If the following results were obtained in an indentation test on annealed copper, derive the Meyer relationship for this material.

Load, P (kg f)	100	300	1000
Impression diameter, d (mm)	1.9	3.0	5.0

$$[P = 21.6d^{2 \cdot 38}]$$

(3) If the Brinell hardness of a structural steel is 420 kgf mm^{-2}, what is the corresponding value of the Meyer hardness if the ratio of indentation diameter to indenter diameter = 0.5 in both cases? Assume that the same indenter load is used in each test.

$$[450.4 \text{ kgf mm}^{-2}]$$

(4) The U.S. standard tension test bar for steels has a relationship given by $\ell \approx 4.51\sqrt{A}$, while the U.K. standard tension bar relationship is given by $\ell \approx 5.65\sqrt{A}$. (A is the cross-sectional area, mm^2, and ℓ the gauge length, mm.) Explain the basis for the choice of these standards and show that the results using both standards are comparable.

(5) A test bar of steel plate 38 mm wide and 16 mm thick when tested in tension gave the following percentage elongations, x, for gauge lengths, ℓ.

ℓ (cm)	10	15	20	25	30	35
x	37.8	31.8	28.5	26.7	25.5	24.5

Using the above data determine the approximate percentage elongation in a gauge length of 20 cm of a bar of the same steel 25 mm wide and 16 mm thick.

$$[26.74\%]$$

(6) A tension member with a diameter of 13 mm and initial gauge length length of 25 mm was used to obtain the true stress-true strain diagram. For loads of 50 kN and 53 kN the total extensions were found to be 2.5 mm and 13 mm, respectively. Determine (a) the value of the strength coefficient k and the strain-hardening exponent, n, assuming that the relationship between true stress, σ_t, and true strain, ε, is given by $\sigma_t = k\varepsilon^n$; (b) the change in gauge length of the specimen at the maximum load.

$$[\sigma_t = 757\varepsilon^{0 \cdot 26} \text{ MN m}^{-2}; 7.42 \text{ mm}]$$

16

(7) From the table of data below estimate the tangent and secant moduli at 3% strain.

Stress (MN m^{-2})	4	8	12	16	20	23
Strain (%)	0.50	1.00	1.65	2.60	3.85	5.10

Explain why the moduli are much less than the elastic modulus of metals.

(London)   [567, 345 MN m^{-2}]

2   DIFFUSION

Diffusion is the mechanism by which matter is transported through matter. Because the movement of each individual atom or particle is obscured by neighbouring atoms or particles, its motion is an apparently aimless series of flights and collisions; the net result however, is an over-all and specific displacement of matter.

The simplest possible diffusion system is shown in figure 2.1, where J_x, the flux of the diffusing species is positive from left to right as the diffusing species moves from an initial high concentration C_s to a lower one C_x over a distance Δx under steady-state conditions. Flux is defined as the amount of material passing through a unit area perpendicular to the flux direction (flux is a vector) per unit time. In this case, C_s and C_x are constant, the concentration gradient dC/dx is constant, and since $C_s > C_x$, the

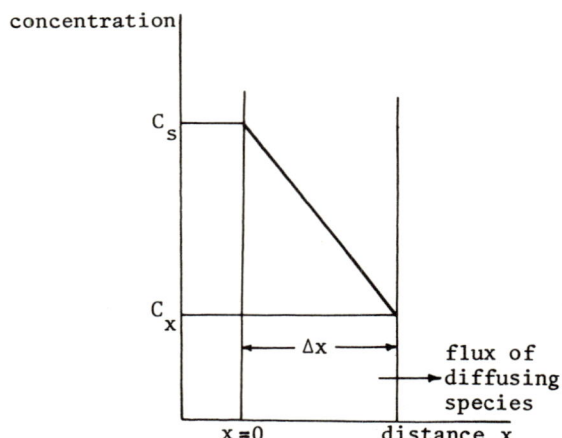

Figure 2.1

concentration gradient is negative from left to right. We can thus write

$$J_x = -D \frac{dC}{dx} \tag{2.1}$$

where D = diffusivity or diffusion coefficient of the diffusing species. Equation 2.1 is known as Fick's first law of diffusion.

A more common instance of diffusion arises when the concentration of the diffusing species, for example C_x in figure 2.1, changes with

18

time. Under these conditions (known as transient or unsteady state conditions) the concentration gradient, $\partial C/\partial x$, and hence the flux, changes as time passes. We can thus write

$$\frac{dC_x}{dt} = \frac{d}{dx}\left(D\,\frac{dC_x}{dx}\right)$$

(2.2)

This is known as Fick's second law of diffusion.

Equations similar to equation 2.2 are encountered frequently in practical problems, for example in heat and mass transfer. Particularly useful in describing many solid state diffusion processes is the case of diffusion in a semi-infinite solid. Here the concentration of the diffusing species, C_x, varies with distance x and time t. (Strictly speaking it should also vary with D but in this analysis it is assumed that it does not.) Then we can write equation 2.2 as

$$\frac{dC_x}{dt} = D\,\frac{d^2C_x}{dx^2}$$

which yields the solution

$$\frac{C_x - C_0}{C_s - C_0} = 1 - \mathrm{erf}\left(\frac{x}{2\sqrt{(Dt)}}\right)$$

or $$\frac{C_s - C_x}{C_s - C_0} = \mathrm{erf}\,\frac{x}{2\sqrt{(Dt)}}$$

$(C = C_0$ at $t = 0$, $0<x<\infty$; $C = C_s$ at $x = 0$, $0<t<\infty$; that is, the surface is maintained at concentration C_s - see figure 2.2.)

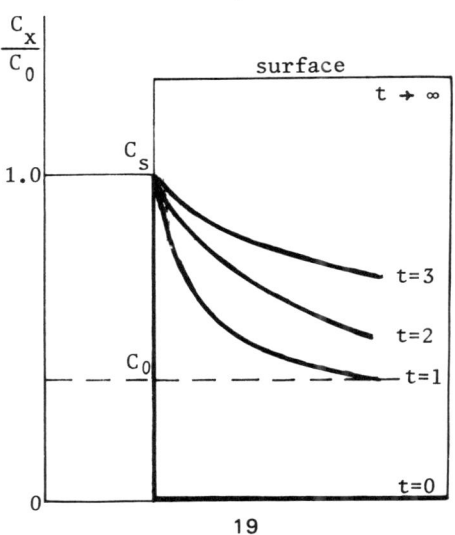

Figure 2.2

19

Erf is the Gaussian error function or normalised probability integral, i.e.

$$\text{erf } y = \frac{2}{\sqrt{\pi}} \int_0^y \exp(-Z^2)\, dZ$$

Values of erf can be obtained from standard tables (appendix II).

Example 2.1

In an experiment to study the diffusion of an interstitial solute in an f.c.c. metal under steady-state conditions, a saturated solution of the solute in the metal was maintained on one face of a piece of metal foil and the other face was kept at zero concentration. The following data were obtained for a foil thickness of 0.25 mm and 1000 mm^2 area.

Temperature (K)	Solubility of Solute in Metal (kg m^{-3})	Rate of Solute Diffusion through Foil (g s^{-1})
1223	14.4	0.0025
1136	19.6	0.0014

Estimate the diffusion coefficients at each of these temperatures and obtain a value for the activation energy for the diffusion of the solute in the f.c.c. metal, assuming Fick's laws to apply. (Nottingham)

Area of foil = 1000 mm^2 = 10^{-3} m^2 so the rates of solute diffusion through the foil at various temperatures become

Temperature (K)	Solubility of Solute in Metal (gm^{-3})	Rate of Solute Diffusion through Foil (gm^{-2} s^{-1})
1223	14 400	2.5
1136	19 600	1.4

By Fick's first law of diffusion

$$\text{rate of solute diffusion, } \dot{m} = \left| D \frac{\Delta C}{\Delta x} \right|$$

where ΔC = (solubility of solute in metal - surface concentration of solute), $\Delta x = 25 \times 10^{-5}$ m. It is assumed that surface concentration of solute is zero so

$$D = \dot{m}\Delta x / (\text{solubility of solute in metal})$$

whence

$$D(1223\text{K}) = \frac{2.5 \times 25 \times 10^{-5}}{14400} = 4.34 \times 10^{-8} \text{ m}^2 \text{ s}^{-1}$$

and $D(1136K) = \dfrac{1.4 \times 25 \times 10^{-5}}{19600} = 1.79 \times 10^{-8}$ m^2 s^{-1}

Assuming diffusivity follows an Arrhenius rate law, we can write

$$D = A \exp\left(-\frac{Q}{RT}\right)$$

i.e. $\dfrac{D(1223K)}{D(1136K)} = \dfrac{4.34 \times 10^{-8}}{1.79 \times 10^{-8}} = \dfrac{\exp[-Q/(8.314 \times 1223)]}{\exp[-Q/(8.314 \times 1136)]}$

whence $Q = 122$ kJ mole^{-1}

Example 2.2

For small values of $y(<0.5)$ the Gaussian error function can be approximated as $\mathrm{erf}(y) \simeq y$. In the case of a piece of silicon exposed to aluminium vapour at 1300 °C, aluminium diffuses into the silicon. Find out how long it will take for the concentration of aluminium at a point 0.01 cm from the surface to be 35% that at the surface. The diffusion coefficient for aluminium in silicon is given in figure 2.3.

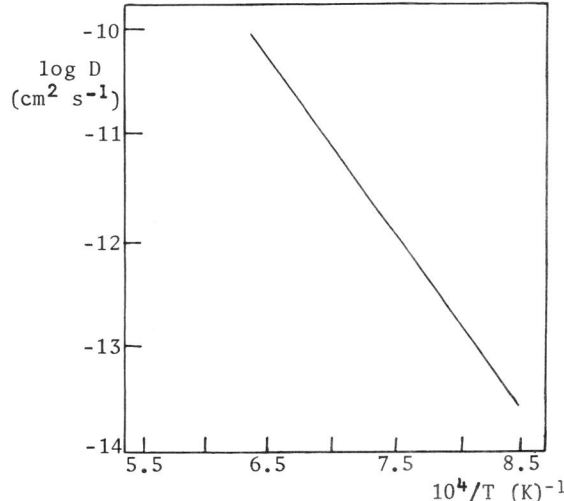

Figure 2.3

At $T = 1573$ K, $D = 10^{-10}$ cm^2 s^{-1} (from figure 2.3). Fick's second law in the approximate form is

$$\frac{C_s - C_x}{C_s - C_0} \simeq \frac{x}{2\sqrt{(Dt)}}$$

With $C_x = 0.35 C_s$, $x = 0.01$ cm, $D = 10^{-10}$ cm^2 s^{-1} and assuming $C_0 = 0$,

21

we have

$$0.65 \simeq \frac{0.01}{2\sqrt{(10^{-10}t)}}$$

whence t = 164 h.

Example 2.3

A foreman in charge of the heat treatment of gears embarks on an economy programme, aimed at increasing the life of his electric furnaces. He therefore suggests reducing the carburising temperature from 1000 to 900 °C because he believes the furnace life is much longer at a lower temperature. He further contends that the carburising time required to obtain the same result will only be about 10% longer.

Assume that at both temperatures the initial and surface concentration and that at 0.025 cm below the surface are 0.2%C, 0.9%C and 0.81%C, respectively. Also assume that diffusivity of carbon bears an Arrhenius-type relationship to carburising temperature, with an activation energy of 144 kJ mole^{-1}.

Comment critically on the contentions of the foreman. (Gas constant = 8.314 J mole^{-1} K^{-1}.)

If diffusivity of carbon, D, bears an Arrhenius-type relationship to carburising temperature, T, then we can write

$$D = D_0 \exp(-Q/RT)$$

where D_0 and Q are constants and R and T have their usual meanings. Thus at the two carburising temperatures we have

$$\ln\left(\frac{D_1}{D_2}\right) = -\frac{Q}{R}\left(\frac{1}{T_1} - \frac{1}{T_2}\right)$$

Putting in the respective values we obtain

$$\ln\left(\frac{D_1}{D_2}\right) = -\frac{144000}{8.314}\left(\frac{1}{1273} - \frac{1}{1173}\right)$$

giving $D_1/D_2 = 3.2$.

Solution of Fick's second law is

$$\frac{C_s - C_x}{C_s - C_0} = \mathrm{erf}\left(\frac{x}{2\sqrt{(Dt)}}\right)$$

At both temperatures the left-hand side of the above equation is identical. Also, the argument of the error function has the same value. Hence

$$\frac{x_1}{2\sqrt{(D_1 t_1)}} = \frac{x_2}{2\sqrt{(D_2 t_2)}}$$

with $x_1 = x_2 = 0.025$ cm, $t_2/t_1 = D_1/D_2 = 3.2$.

Thus the contention of the foreman regarding the ratio of the carburising times is incorrect. For the same result at 900 °C it takes 220% more time (not 10%) as at 1000 °C.

Example 2.4

The diffusivity, D, of carbon during case hardening is defined by

$$D = A \exp(-Q/RT)$$

where A = constant, Q = activation energy, R = gas constant, T = absolute temperature. The concentration of carbon in this case obeys Fick's second law of diffusion, given as

$$\frac{C_s - C_x}{C_s - C_0} = erf\left(\frac{x}{2\sqrt{(Dt)}}\right)$$

where C_s, C_x and C_0 are the concentrations at the surface, at depth x and at the start of the process, respectively, and t is time in seconds.

Calculate the concentration of carbon at a depth of 0.2 mm in a 0.2%C mild steel which has been case hardened for 10 h at 850 °C. Assume the surface concentration of carbon is 0.9%. ($A = 25$ mm^2 s^{-1}; $Q = 145$ kJ mole^{-1}; $R = 8.314$ J mole^{-1} K^{-1})

$$\text{Diffusivity, } D = A \exp\left(-\frac{Q}{RT}\right)$$

With the numerical values of the parameters, we have

$$D = 25 \exp\left(\frac{-145 \times 10^3}{8.314 \times 1123}\right) \text{ mm}^2 \text{ s}^{-1} = 4.5 \times 10^{-6} \text{ mm}^2 \text{ s}^{-1}$$

Applying Fick's second law, we have

$$\frac{0.9 - C_x}{0.9 - 0.2} = erf\left(\frac{0.2}{2\sqrt{(4.5 \times 10^{-6} \times 36\ 000)}}\right) = erf(0.248)$$

whence $C_x = 0.745\%$C.

Example 2.5

In the manufacture of parts, such as gears, a hardened surface layer

23

and a tough structure in the interior are required. The first step in the process is the diffusion of carbon into the steel surface thereby increasing the level from about 0.2%C (the initial concentration) to 0.5 to 0.9%C for 0.0127 to 0.127 cm.

When such a part is placed in a furnace at 1000 °C with an environment rich in hydrocarbon gases, the surface carbon content of about 0.9% is attained very quickly. The carbon content will then rise gradually beneath the surface as a function of time.

Calculate the content at 0.025 cm below the surface after 10 h at 1000 °C. (Diffusivity of carbon in steel is 0.3×10^{-6} cm^2 s^{-1}.)

Fick's second law, with the symbols having their usual meanings, is

$$\frac{C_s - C_x}{C_s - C_0} = \text{erf}\left(\frac{x}{2\sqrt{(Dt)}}\right)$$

Then $\dfrac{0.9 - C_x}{0.9 - 0.2} = \text{erf}\left(\dfrac{0.025}{2\sqrt{(0.3 \times 10^{-6} \times 36\ 000)}}\right) = \text{erf}(0.120)$

or $\dfrac{0.9 - C_x}{0.7} = 0.135$

whence $C_x = 0.806\%C$. The carbon content at 0.025 cm below the surface is 0.81%.

Example 2.6

A solution to Fick's second law of diffusion is

$$\frac{C_x - C_b}{C_s - C_b} = 1 - \text{erf}\left(\frac{x}{2\sqrt{(Dt)}}\right)$$

where C_x, C_b, C_s are the concentrations at depth x, in the bulk and at the surface, respectively, and the other symbols have their usual meanings. Under certain conditions the diffusion coefficient D for a metal A entering a metal B can be taken as 6×10^{-7} cm^2 s^{-1}.

Calculate the time required to diffuse A, to a concentration of 1.5% at a depth of 0.5 mm, in a specimen of B initially containing 0.1%A, the concentration of A at the surface being maintained at 10%. (A table of error functions is provided in appendix II.)

For the given concentrations

$$\text{erf } X = 1 - \left(\frac{1.5 - 0.1}{10 - 0.1}\right) = 0.86$$

where $X = x/[\sqrt{(Dt)}/0.5]$. Using the error function table, we have

$$\frac{x}{2\sqrt{(Dt)}} = 1.06$$

Inserting the relevant numerical values, we have

$$\sqrt{t} = \frac{0.5}{2.12\sqrt{(6 \times 10^{-5})}}$$

whence $t = 15.44$ min $= 0.25$ h, thus the required diffusion time is ¼ h.

Example 2.7

The planar surface of a 0.7%C steel is being decarburised at 1200 K under conditions which maintain a constant carbon concentration of 0.1% at the surface. After time t the concentration at a distance 1 mm from the surface was 0.4%. How much longer would be required for the carbon to attain 0.4% concentration at a depth of 2.5 mm? (The mean diffusion coefficient, D_c, for carbon in austenite is 2×10^{-5} mm^2 s^{-1}.)

After time t_1, Fick's law in its approximate form can be written as

$$\frac{0.1 - 0.4}{0.1 - 0.7} = \frac{1.0}{2\sqrt{(Dt_1)}} \tag{1}$$

After time t_2, Fick's law gives

$$\frac{0.1 - 0.4}{0.1 - 0.7} = \frac{2.5}{2\sqrt{(Dt_2)}} \tag{2}$$

From equations 1 and 2

$$t_2 = 6.25\, t_1$$

Thus increase in time is $5.25t_1$.

To obtain time t_1, we have

$$\frac{0.1 - 0.4}{0.1 - 0.7} = \frac{1.0}{2\sqrt{(2 \times 10^{-5}t_1)}}$$

or $t_1 = 13.9$ h

whence increase in time $= 73$ h.

Example 2.8

In certain cases diffusion calculations can be used to estimate the time required to homogenise a material. In such circumstances, the simplest approach is to estimate the time required to reduce the difference in concentration to one-half. (a) Assuming that the initial concentration is zero, show that the homogenisation time is approximately x^2/D, where x is some depth and D the diffusivity. (b) Calculate the ratio of the homogenisation times for nickel and carbon in γ-iron, given that log D for C in γ-iron and log D for Ni in γ-iron are -6.5 and -11.5, respectively.

(a) Fick's second law is

$$\frac{C_s - C_x}{C_s - C_0} = \text{erf} \left(\frac{x}{2\sqrt{(Dt)}}\right)$$

with $C_0 = 0$ and $C_x = 0.5C_s$ we have

$$0.5 = \text{erf}\left[\frac{x}{2\sqrt{(Dt)}}\right]$$

or $\quad \dfrac{x}{2\sqrt{(Dt)}} = 0.478$

i.e. $\dfrac{x^2}{Dt} = 0.914$

or $\quad t \simeq \dfrac{x^2}{D}$

(b) $\quad \dfrac{t(\text{Ni in γ-iron})}{t(\text{C in γ-iron})} \simeq \dfrac{x^2}{D(\text{Ni in γ-iron})} \times \dfrac{D(\text{C in γ-iron})}{x^2}$

$$\simeq \frac{10^{-6 \cdot 5}}{10^{-11 \cdot 5}} \simeq 10^5$$

Example 2.9

In order to avoid coarsening of the grains it is advisable to perform the case hardening of steel at 850 instead of 900 °C. Calculate the diffusion coefficient, D, of carbon in austenite for the two temperatures using the formula

$$D = D_0 \exp(-Q/RT)$$

The values of D_0 and Q for the system are 0.21 cm^2 s^{-1} and 140 140 kJ mole^{-1}; R, the gas constant = 8.314 J mole^{-1} K^{-1}.

What time is necessary to case harden at 850 °C if it is required to get the same result as is obtained when case hardening at 900 °C for 10 h?

The diffusion coefficients at 850 and 900 °C are, respectively

$$D_1 = 0.21 \exp\left(\frac{-140 \times 10^3}{8.314 \times 1123}\right) = 646 \times 10^{-10} \ cm^2 \ s^{-1}$$

and $D_2 = 0.21 \times \exp\left(\frac{-140 \times 10^3}{8.314 \times 1173}\right) = 1224 \times 10^{-10} \ cm^2 \ s^{-1}$

Equating the error function part of Fick's second law in its approximate form, we have

$$\frac{x_1}{2\sqrt{(D_1 t_1)}} = \frac{x_2}{2\sqrt{(D_2 t_2)}}$$

thus (for $X_1 = X_2$)

$$t_1 = \frac{D_2 t_2}{D_1} = \frac{1224 \times 10^{-10} \times 10}{646 \times 10^{-10}} \ h$$

whence required time at 850 °C = 19 h.

Example 2.10

The concentration of hydrogen in α-iron is given by the expression

$$C = 42.7 \ p^{\frac{1}{2}} \exp\left(-\frac{6500}{RT}\right)$$

where C is in parts per million by weight and p is the external hydrogen pressure in units of atmosphere; R and T have their usual meanings.

The diffusivity, D, of hydrogen in α-iron is given by

$$D = 1.4 \times 10^{-3} \exp\left(-\frac{3200}{RT}\right) \ cm^{-2} \ s^{-1}$$

An iron membrane, of thickness 10^{-2} cm, separates a reservoir of hydrogen at a pressure of 200 atm from another one at a pressure of 2 atm at 200 °C. Calculate the flux of hydrogen through the membrane.

At pressures of 200 and 2 atm, respectively, the concentrations of hydrogen in α-iron are

$$C_1 = 42.7 \times 200^{\frac{1}{2}} \exp\left(-\frac{6500}{8.314 \times 473}\right) = 116 \ ppm$$

and $C_2 = 42.7 \times 2^{\frac{1}{2}} \exp\left(-\frac{6500}{8.314 \times 473}\right) = 11.6 \ ppm$

thus $\Delta C = C_1 - C_2 = 104.4$ ppm

From the relationship for diffusivity, we calculate it at T = 473 K to be

$$D = 1.4 \times 10^{-3} \exp\left(-\frac{3200}{8.314 \times 473}\right) = 0.62 \times 10^{-3} \ cm^2 \ s^{-1}$$

Fick's first law states

$$J = D\left|\frac{\Delta C}{x}\right|$$

thus flux, $J = \dfrac{0.62 \times 10^{-3} \times 104.4}{10^{-2}} = 6.4$ cm s^{-1} ppm

Example 2.11

Copper oxidises to cuprous oxide, exhibiting parabolic behaviour at temperatures above 500 °C. The activation energy for the process is 158 kJ mole^{-1} which corresponds to the activation energy for the diffusion of cuprous ions in cuprous oxide.

If 10^{-3} cm of cuprous oxide forms after exposure to air for 5 min at 600 °C, calculate the oxide thickness when a clean copper surface is exposed to air at 550 °C for 10 min.

The difference in oxygen concentration between oxide-oxidant and metal-oxide interfaces remains constant during the oxidation process and is designated ΔC. J_0, the oxygen flux, is, from Fick's first law

$$J_0 = -D_0\frac{\Delta C}{x} \tag{1}$$

where D_0 = diffusivity of oxygen in the oxide of thickness x.

The rate of increase of oxide thickness, dx/dt, is proportional to the flux of oxygen atoms reaching the metal-oxide interface, so that

$$\frac{dx}{dt} = kJ_0$$

where k is a constant of proportionality and t is time.

$$\frac{dx}{dt} = -kD_0\frac{\Delta C}{x}$$

which on integration gives

$$x^2 = AtD_0 \tag{2}$$

where $A = -2k\Delta C$.

Diffusivity is related in an Arrhenius fashion to temperature, T, so that

$$x^2 = A' \exp(-Q/RT)t \tag{3}$$

where A' is a constant, and Q, R and T have their usual meanings. Putting the relevant numerical values for the parameters in the equation 3 for the two cases we have

28

$$\frac{(10^{-3})^2}{\exp\left(-\dfrac{158\times10^3}{8.314\times873}\right)\times5}=\frac{x^2}{\exp\left(-\dfrac{158\times10^3}{8.314\times823}\right)\times10}$$

whence oxide thickness = 7.5×10^{-4} cm.

PROBLEMS

(1) Pieces of a 0.6%C steel have been decarburised at 930 °C in an atmosphere which maintains the surface carbon concentration at 0%. At this temperature the steel was held for 10^4 s. Estimate the depth of the 0.3%C level if the diffusion coefficient, D, for carbon in γ-iron is given by the equation

$$D = 0.01\ \exp(-147000/RT)\ \text{cm}^2\ \text{s}^{-1}$$

[0.006 cm]

(2) Carburisation of a pure iron is carried out at 950 °C. It is desirable to achieve a carbon content of 0.9% at a depth of 0.01 cm, assuming that the carbon content of the surface is maintained at 1.2% and the diffusion coefficient for carbon in γ-iron is 10^{-6} cm^2 s^{-1}. Calculate the time necessary for the process.

[440 s]

(3) A new steel being marketed for automotive applications consists of a low alloy steel core and a high chromium surface layer diffused into the steel for increased corrosion resistance. Calculate how long a heat treatment would be required at 1000 °C to diffuse chromium far enough into the steel so that the composition is 18%Cr at 0.02 cm below the surface and 100%Cr at the surface. Assume the pre-exponential constant in the expression for the diffusion coefficient for chromium in steel to be 0.47 cm^2 s^{-1} and the activation energy for the diffusion process to be 333 kJ mole^{-1}.

[10^{10} s]

(4) The diffusion coefficient for copper atoms in aluminium is found to be 1.28×10^{-22} m^2 s^{-1} at 400 K and 5.75×10^{-19} m^2 s^{-1} at 500 K. Find the temperature at which its value is 10^{-16} m^2 s^{-1}.

Copper atoms diffuse under steady-state conditions at room temperature through 0.1 mm thick aluminium foil. If the concentration of copper atoms is maintained at 10^{29} m^{-3} on one side of the foil and negligible on the other, what is the mass rate of flow of copper through the foil?

[590 K; 4.5×10^{-9} kg m^{-2} s^{-1}]

(5) A diffusion couple is made of a block of pure copper and a block of copper containing 5 at.% radioactive copper isotope. Diffusion proceeds for 35 h at a temperature of 1200 K. At the end of the period, it is found that the concentration of radioactive copper is 0.645 at.% at a plane located 0.013 cm to the right of the interface. Calculate the diffusion coefficient.

[5.3×10^{-10} cm^2 s^{-1}]

29

(6) A piece of pure iron is to be nitrided until the concentration of nitrogen is 0.1 at.% on a plane 0.05 cm below the surface. The process to be used is such that the surface concentration will remain fixed at 0.5 at.% nitrogen throughout the process. Calculate the time for diffusion to be allowed to occur. Diffusion coefficient for nitrogen in pure iron = 3.48×10^{-8} cm^2 s^{-1}.

[6.02 h]

3   VISCOELASTICITY

3.1 INTRODUCTION

The different modes by which materials and especially high polymers
may deform makes a description of their mechanical properties
difficult.   This is particularly true of the conditions in which
different time-dependent mechanisms operate simultaneously.

   Elastic solids and viscous fluids differ widely in their deformational characteristics.   Elastically deformed bodies return to a
natural or undeformed state on removal of applied loads.   Viscous
fluids, however, possess no tendency for deformational recovery.
Elastic stress is directly related to deformation whereas stress in
a viscous fluid depends (except for the hydrostatic component) on
the rate of deformation.   A material that possesses both elastic and
viscous properties is called a viscoelastic material.

   The elastic (Hookean) solid and the viscous (Newtonian) fluid
represent opposite ends of a wide spectrum of viscoelastic behaviour.
Although viscoelastic materials are temperature-sensitive, the
analysis given in this section is isothermal.

3.2 MECHANICAL MODELS

Simple mechanical models may be used in various combinations to
represent the time-dependent stress-strain characteristics of these
viscoelastic materials.   The requirement for these rheological
models is that they exhibit the same relationships between force,
elongation and time, as stress, strain and time do in the real
viscoelastic material.   First, let us consider the two basic components of such rheological models or mechanical analogues: the
linear spring (representing the elastic solid) and the dashpot (representing the viscous fluid).   Let σ be the applied stress, ε the
strain of the linear spring, E the modulus of elasticity, η the
viscosity of the fluid in the dashpot and t the time.   Then for the
linear spring in figure 3.1

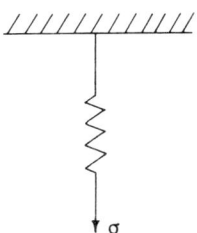

Figure 3.1

$$\sigma = E\epsilon$$

(The element is Hookean and the spring is assumed to have unit cross-sectional area and unit length.)

For the dashpot in figure 3.2

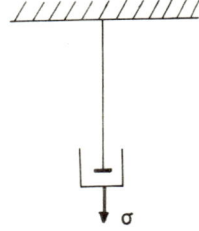

Figure 3.2

$$\sigma = \eta\frac{d\epsilon}{dt}$$

(The element is Newtonian and the dashpot is assumed to have unit cross-sectional area and unit length.)

Maxwell Model

This is a combination of a spring and a dashpot in series (figure 3.3). If ϵ_1 is the strain in the spring and ϵ_2 is that in the dashpot then

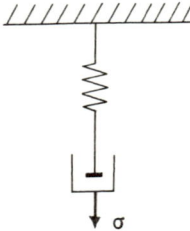

Figure 3.3

$$\text{total strain, } \epsilon_T = \epsilon_1 + \epsilon_2 \tag{3.1}$$

but $\quad \epsilon_1 = \dfrac{\sigma}{E}$

and $\quad \dfrac{d\epsilon_2}{dt} = \dfrac{\sigma}{\eta}$

So differentiating equation 3.1 we have

$$\frac{d\varepsilon_T}{dt} = \frac{d\varepsilon_1}{dt} + \frac{d\varepsilon_2}{dt}$$

and making the substitution from above, gives

$$\frac{d\varepsilon_T}{dt} = \frac{1}{E}\frac{d\sigma}{dt} + \frac{\sigma}{\eta} \tag{3.2}$$

This is the constitutive equation for a viscoelastic material described by the Maxwell model.

Equation 3.2 could also be arrived at from considerations of shear strain. Thus total shear strain, γ_T = shear strain in the spring, γ_1, plus shear strain in the dashpot, γ_2, i.e.

$$\gamma_T = \gamma_1 + \gamma_2 \tag{3.3}$$

$$\frac{d\gamma_T}{dt} = \frac{d\gamma_1}{dt} + \frac{d\gamma_2}{dt}$$

Now $\gamma_1 = \frac{\tau}{G}$

and $\frac{d\gamma_2}{dt} = \frac{\tau}{\eta}$

where τ = shear stress, G = modulus of rigidity of the material. Thus equation 3.3 becomes

$$\frac{d\gamma}{dt} = \frac{1}{G}\frac{d\tau}{dt} + \frac{\tau}{\eta} \tag{3.4}$$

Kelvin-Voight Model

This is a model with a spring and a dashpot connected in a parallel combination and describes well some of the features of creep in thermoplastics (figure 3.4). The spring might be considered to

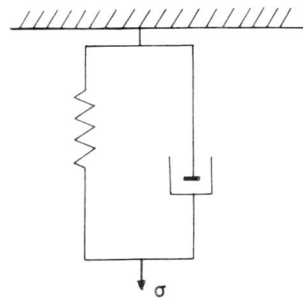

Figure 3.4

33

represent the extension of primary bonds and angles or the entropy elasticity of randomly kinked molecules, while the dashpot represents the time-dependent sliding of main chains and flipping of side chains.

Total stress = stress in spring + stress in dashpot. That is

$$\sigma = \sigma_1 + \sigma_2$$

Now $\sigma_1 = E\epsilon$ and $\sigma_2 = \eta \, d\epsilon/dt$, so

$$\sigma = E\epsilon + \eta\frac{d\epsilon}{dt}$$

$$\sigma - E\epsilon = \eta\frac{d\epsilon}{dt}$$

$$\frac{dt}{\eta} = \frac{d\epsilon}{\sigma - E\epsilon}$$

Integrating gives

$$\frac{t}{\eta} = -\frac{1}{E}\ln(\sigma - E\epsilon) + \text{constant}$$

Example 3.1

A polymer is subjected to a shear stress of 1 MN m^{-2} and the resulting shear strain was found to vary with time in the following manner.

Time (ks)	3.6	7.2	36.0	72.0
Shear strain	0.0060	0.0084	0.0100	0.0100

Assuming the polymer to behave according to the Kelvin-Voight model, calculate its shear modulus and viscosity.
(Nottingham)

For the Kelvin-Voight model, we can write

$$\eta\frac{d\gamma}{dt} + G\gamma = \tau$$

$$\frac{d\gamma}{dt} = \frac{\tau}{\eta} - \frac{G\gamma}{\eta}$$

Thus a plot of $d\gamma/dt$ against γ would give τ/η as the intercept and $-G/\eta$ as the slope.

The data are now calculated in a suitable form for plotting, as follows.

$\dfrac{d\gamma}{dt} \times 10^7 (s^{-1})$	16.67	11.67	2.78	1.39
$\gamma \times 10^4$	60	84	100	100

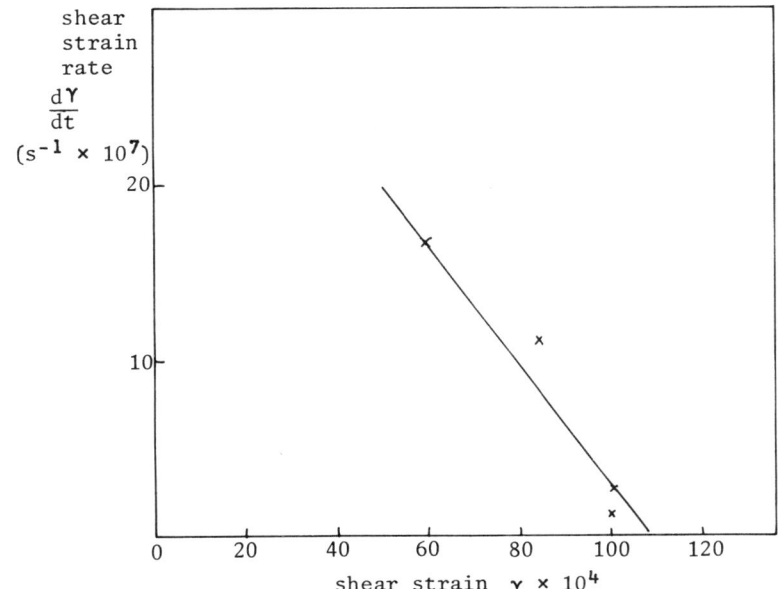

Fig. 3.5

The plot of $d\gamma/dt$ against γ (figure 3.5) gives

   intercept = 37×10^{-7} s^{-1}

but $\tau = 10^6$ N m^{-2} therefore

   $\eta = 27 \times 10^{10}$ N s m^{-2}

   slope = -0.34×10^{-3} s^{-1}

whence $G = 92$ MN m^{-2}.

Example 3.2

A polymer is creep tested in shear with an applied stress of
0.5 MN m^{-2} and the following shear strains were noted.

Time, t (h)	0	5	10	20	30	40	50
Shear strain, γ	0	0.0058	0.0090	0.0128	0.0145	0.0150	0.0150

Stating clearly any assumptions made, calculate the relaxation time
for flow in the polymer, as well as its shear modulus.

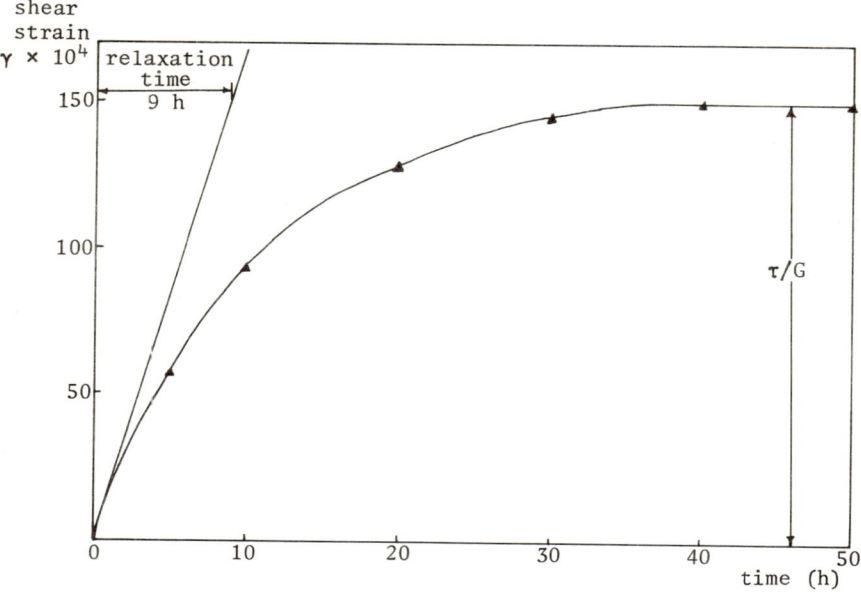

Fig. 3.6

Assume that flow in the polymer is of the Kelvin-Voight type. A plot of γ against t (figure 3.6) straightway gives the relaxation time, T_{rel} = 9 h and the shear modulus, from the fact that

$$0.0150 = \frac{0.5}{G}$$

whence G = 33.3 MN m^{-2}.

Example 3.3

A stress of 310 MN m^{-2} is applied to a piece of glass straining it 0.2%. After 1 year the stress relaxes to 207 MN m^{-2}, maintaining the same strain on the glass. If the glass is maintained in this stretched position for 2 years, to what value will the stress have fallen? (Assume glass is represented by a Maxwell viscoelastic model.)

For the Maxwell model

$$\frac{d\varepsilon}{dt} = \frac{1}{E}\frac{d\sigma}{dt} + \frac{\sigma}{\eta}$$

Strain rate is zero so we have after integration

$$\ln \frac{\sigma}{\sigma_0} = -\frac{t}{\lambda}$$

36

where σ_0 is the stress at time = λ. We can now evaluate λ by saying

$$\ell n \left(\frac{207}{310}\right) = -\frac{1}{\lambda}$$

whence λ = 2.5 years. So that at t = 2 years

$$\sigma = \sigma_0 \exp\left(-\frac{2}{2.5}\right) = 310 \exp\left(-\frac{2}{2.5}\right) = 139 \text{ MN m}^{-2}$$

Example 3.4

A mortar cylinder 114 mm in diameter and 225 mm long is subjected to a stress of 1.4 MN m^{-2}. Assuming the behaviour is as the Maxwell model, calculate (a) the instantaneous deformation; (b) the total deformation after 10 h; and (c) if, after 10 h the strain is kept constant, after what additional time will the stress in the cylinder decay to 80% of its initial value? (E = 5.5 GN m^{-2}; η = 3 × 10^{14} N s m^{-2})

For the Maxwell model

$$\frac{d\varepsilon_T}{dt} - \frac{1}{E}\frac{d\sigma}{dt} - \frac{\sigma}{\eta} = 0 \tag{1}$$

(a) Instantaneous deformation $\delta_{in} = \frac{\sigma}{E}$ × original length

$$= \frac{1.4 \times 10^6}{5.5 \times 10^9} \times 225 = 57.27 \times 10^{-3} \text{ mm}$$

(b) After time t, deformation is now

$$\delta_T = \frac{\sigma}{E} \times \text{original length} + \frac{\sigma t}{\eta} \times \text{original length}$$

$$= \left(57.27 \times 10^{-3} + \frac{1.4 \times 10^6 \times 36\,000 \times 225}{3 \times 10^{14}}\right)$$

$$= 95.07 \times 10^{-3} \text{ mm}$$

(c) Putting $d\varepsilon_T/dt$ = 0 in equation 1 and integrating, we obtain

$$\frac{\eta}{E}\int_2^1 \frac{d\sigma}{\sigma} = \int_1^2 dt$$

With the values of η, E and σ, we write

$$\frac{3 \times 10^{14}}{5.5 \times 10^9} \times \ell n\left(\frac{10}{8}\right) = (t_2 - t_1) = \Delta t$$

37

whence Δt = 3 h 23 min, the additional time needed.

Example 3.5

A tensile stress of 10 MN m^{-2} is required to deform a piece of rubber by 0.5 mm per mm. After this strain has been maintained for 50 days the stress is found to have decreased to 5 MN m^{-2}. Calculate the stress required to maintain the same strain after a total time of 100 days.

   With time, stresses are relaxed in those situations where they are initially developed from elastic elongation. The time required for the adjustment of stresses is known as the relaxation time, and is defined as the time it takes for the stress to be reduced to its original value divided by e. Thus under conditions of constant strain, we can write

$$\frac{d\sigma}{dt} = - \frac{\sigma}{\lambda}$$

$$\sigma = \sigma_0 \exp(-t/\lambda)$$

where σ_0 is the original stress and λ the relaxation time. In this example, we can write

$$\ell n \left(\frac{5}{10}\right) = - \frac{50}{\lambda}$$

whence λ = 72.1 days. Thus the stress after 100 days, σ_{100}, is

$$\sigma_{100} = 10 \exp\left(- \frac{100}{72.1}\right) = 2.5 \text{ MN m}^{-2}$$

PROBLEMS

(1) (a) Show that for a Maxwell viscoelastic model in a constant strain condition with initial stress, σ_0, the stress at time t is given by

$$\sigma = \sigma_0 \exp(-t/\lambda)$$

   (b) Show that for a Kelvin viscoelastic model of elastic modulus, E, in creep conditions, the strain is given by

$$e = \frac{\sigma}{E}[1 - \exp(-t/\lambda)]$$

Explain the nature of the non-dimensional quantity, t/λ, in the above equations, making reference to the properties of named polymers.

(2) A plastic material which may be represented by a Maxwell viscoelastic model has an instantaneous tensile modulus of 5 MN m^{-2}. In a constant strain experiment the stress falls to 50% of its initial

value after 10^6 s. Estimate the value of the relaxation time for this material.

[401 h]

(3) Describe the time-dependent mecahnical properties of a material which can be represented by a model consisting of a dashpot and a spring connected in series. Sections of a water pump casing are held together by nylon bolts which are tightened until they are under a tensile stress of 50 MN m^{-2}. The water pressure is such that the joints will leak if the stress in the bolts falls to 40 MN m^{-2}. How long will the joint survive without leaking if the bolts have a modulus of elasticity of 1.5 GN m^{-2} and an effective viscosity at the temperature of operation of 7.5×10^{17} N m^{-2} s?
(Nottingham)

[31 000 h]

(4) A metal spring is stretched at 35 MN m^{-2} at high temperature. After 1000 h, it is found that the stress in the stretched spring is only 28 MN m^{-2}. Calculate the relaxation time. What stress would remain in the spring after 2000 h?

[4482 h; 22.4 MN m^{-2}]

(5) A steel bolt joining two flanges is stressed initially up to 207 MN m^{-2}. What will be the stress in the bolt after 144 h if it is to operate at a temperature of 510 °C, given that the relaxation time for the stress at this temperature is 207 h?

[103 MN m^{-2}]

4   COMPOSITE MATERIALS

4.1 INTRODUCTION

Usually most high-strength materials possess poor ductility. Such
materials are extremely notch-sensitive and to attain their full
tensile properties demand an extremely high surface finish, making
them unsuitable for most practical situations. One method of using
such materials would be to encase them in a softer, more ductile
material. This is the basis of the composite material, in which the
high-strength material, usually in the form of fibres, is surrounded
by a cylinder or matrix of much lower modulus material.

Many common structural materials are composites of one form or the
other. Familiar examples are steels containing mixed microstructures,
concrete (sand, gravel and hydrated cement), fibreglass (glass fibres
in a polymer matrix) and wood (cellulose and lignin).

Characteristics of some fibres used in composite materials

Fibre	Modulus of Elasticity (GN m^{-2})	Density (kg m^{-3} 10^{-3})
E glass	72.4	2.55
S glass	84.1	2.49
Asbestos	160.0	2.50 - 3.40
Carbon	415.0	2.0
Silicon	72.0	2.5

4.2 MECHANICS

When the composite material is stretched, the fibres and matrix
stretch together and the stress in the fibre increases much more than
that in the matrix and most of the stress is transmitted via the
fibres. There are two cases of the deformation of composite materials
to consider.

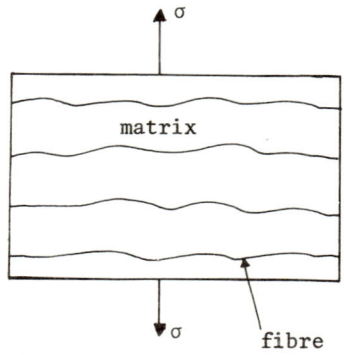

Figure 4.1

40

Case 1 - Isostress

Suppose that a stress σ is applied transverse to the fibres or laminates, such that the laminates are bonded in series (figure 4.1).

Let V_f be the volume of the fibres so that $(1 - V_f)$ is the volume of the matrix. Then the modulus of elasticity of the composite material, E_c, can be computed by dividing the applied stress by the total strain suffered by the material, e_T given by

$$e_T = \frac{\sigma}{E_f}V_f + \frac{\sigma}{E_m}(1 - V_f)$$

E_f and E_m are the moduli of elasticity of the fibre and the matrix materials, respectively. Since

$$E_c = \frac{\sigma}{e_T}$$

we have

$$E_c = \frac{E_f E_m}{V_f E_m + (1 - V_f)E_f}$$

Case 2 - Isostrain

Alternatively, suppose that the stress is applied parallel to the fibres (figure 4.2). If the fibres are well bonded together so that the strains in all fibres are the same, the elastic stresses in the fibres will vary. The total stress, σ_T, is given by the sum of the loads carried by each phase

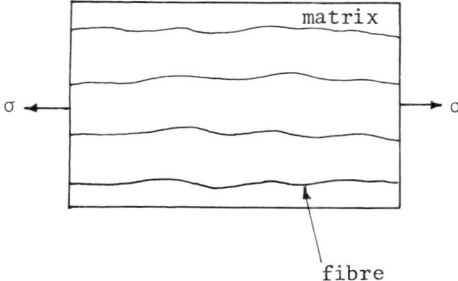

Figure 4.2

$$\sigma_T = E_f e V_f + E_m e (1 - V_f)$$

so that

$$E_c = \frac{\sigma_T}{e} = E_f V_f + E_m (1 - V_f)$$

Example 4.1

A composite material is formed using 1 kg of long unidirectional glass fibres (tensile modulus 85 GN m^{-2}, density 2500 kg m^{-3}) embedded in 5 kg epoxy resin (tensile modulus 5 GN m^{-2}, density 1200 kg m^{-3}). Evaluate the tensile modulus of the fully cured composite (a) in the fibre direction and (b) in the transverse direction.

Let subscripts f, r and c refer to the fibre, resin and composite material respectively, and let subscripts p and t refer to the parallel and transverse directions respectively.

(a) Parallel with fibres

$$e_f = e_r = e_c$$

σ is shared

$$\sigma_c = V_f \sigma_f + (1 - V_f) \sigma_f$$

$$E_p = \frac{\sigma_c}{e_c} = V_f E_f + (1 - V_f) E_r$$

(b) Transverse direction

$$\sigma_f = \sigma_c = \sigma_r$$

$$e_c = e_f(V_f) + (1 + V_f) e_r$$

$$E = \frac{\sigma_c}{e_c} = \frac{\sigma}{e_r + V_f(e_f - e_r)}$$

$$= \frac{E_r}{1 + (V_f e_f / e_r) - V_f}$$

But $E_r = \sigma/e_r$ and $E_f = \sigma/e_f$, therefore

$$e_r E_r = e_f E_f$$

42

or $\quad \dfrac{e_f}{e_r} = \dfrac{E_r}{E_r}$

thus $E_t = \dfrac{E_r}{1 + V_f[(E_r/E_f) - 1]}$

where V_f = volume fraction of the fibres. Now 1 g of fibre occupies 0.4 cm³ and 5 g of resin occupy 4.167 cm³, therefore

$$V_f = \dfrac{0.4}{4.567} = 0.0876$$

thus $E_p = 0.0876(85) + 0.912(5) = 12 \quad$ GN m^{-2}

and $\quad E_t = \dfrac{5}{1 + 0.0876[(1/17) - 1]} = 5.45$ GN m^{-2}

Example 4.2

A tubular shaft of 50 mm external diameter, 40 mm internal diameter and 300 mm length is to be fabricated from freshly drawn glass fibres in a matrix of polyester resin with 50% by volume of fibres. The shaft is to be subjected only to pure torsion in either direction. Estimate the torsional stiffness of the shaft, assuming Young's moduli for glass and polyester resin to be 70 GN m^{-2} and 3.5 GN m^{-2}, respectively. Take Poisson's ratio of the composite material to be 0.2.

State clearly any assumptions made.
(Cambridge)

Assuming the stress is parallel to the fibres, the composite material modulus of elasticity is given as

$$E_{c_p} = V_f E_f + (1 - V_f) E_r$$

$$= 0.50 \times 70 + 0.50 \times 3.5$$

$$= 36.75 \text{ GN m}^{-2}$$

Assuming the stress is transverse to the fibres, then the composite material modulus of elasticity is now

$$E_{c_t} = \dfrac{E_r}{1 + V_f[(E_r/E_f) - 1]}$$

$$= \dfrac{3.5}{1 + 0.5[(3.5/70) - 1]} = 6.67 \text{ GN m}^{-2}$$

Now for the composite material, $E_c = 2G_c(1 + \gamma_c)$, thus

$$G_c = E_c/2.4$$

So for parallel aligned fibres, $G_{c_p} = 15.31 \text{ GN m}^{-2}$ and for transversely aligned fibres, $G_{c_t} = 2.78 \text{ GN m}^{-2}$.

From simple torsion theory, torsional stiffness ($= T/\theta$) is GJ/ℓ, so for parallel aligned fibres

$$\text{torsional stiffness} = \frac{15.31 \times 10^9 \times 36.23 \times 10^{-8}}{0.3}$$

$$= 18.5 \text{ KN m rad}^{-1}$$

and for transversely aligned fibres

$$\text{torsional stiffness} = \frac{2.78 \times 10^9 \times 36.23 \times 10^{-8}}{0.3}$$

$$= 3.36 \text{ KN m rad}^{-1}$$

Example 4.3

For a fibre-reinforced composite with the fibres aligned in the direction of the maximum tensile stress, (a) sketch a graph to show how its strength varies with volume fraction of the fibres; (b) show that there is a minimum volume fraction of fibres, $V_{f\ min}$, corresponding to a minimum strength of the composite and a critical volume fraction of fibres, $V_{f\ crit}$, necessary to give reinforcement, given by

$$V_{f\ min} = \frac{\sigma_{M\ max} - \sigma_{M\ y}}{\sigma_{f\ max} + \sigma_{M\ max} - \sigma_{M\ y}}$$

$$V_{f\ crit} = \frac{\sigma_{M\ max} - \sigma_{M\ y}}{\sigma_{f\ max} - \sigma_{M\ y}}$$

where $\sigma_{M\ max}$ = matrix ultimate tensile strength, $\sigma_{f\ max}$ = fibre ultimate tensile strength, $\sigma_{M\ y}$ = matrix yield strength.

(a) In this case the average stress σ_c acts on a cross-sectional area A_c, σ_f acts on the cross-sectional area A_f, and σ_M acts on the cross-sectional area A_M. Thus the resultant force on the element of composite materials is

$$P = \sigma_c A_c = \sigma_f A_f + \sigma_M A_M$$

The volume fractions of fibres and matrix are $V_f = A_f/A_c$ and $V_M = A_M/A_c$; thus

$$\sigma_c = \sigma_f V_f + \sigma_M V_M$$

The plot of the variation of composite strength with fibre volume fraction is shown in figure 4.3.

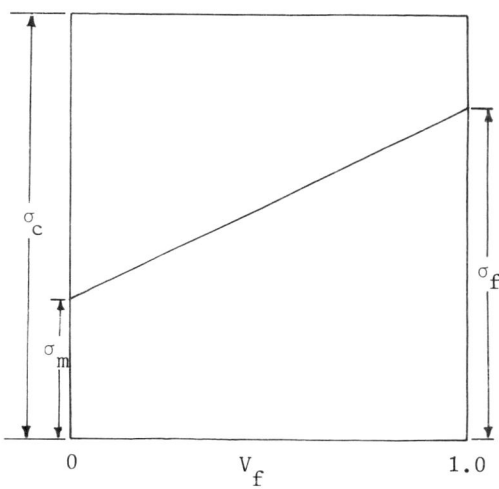

Figure 4.3

(b) If the composite has more than a certain minimum volume fraction of fibres the ultimate strength is achieved when all the fibres are strained to correspond to their maximum (ultimate) stress. In terms of strain, we can write

$$e_{c\ max} = e_{f\ max}$$

The schematic stress-strain curves for the fibres and matrix (figure 4.4) aid in the reasoning to obtain the composite strength. Thus if the fibre strain is assumed equal to the matrix strain in the direction of the fibres, the ultimate strength of the composite is

$$\sigma_{c\ max} = \sigma_{f\ max} V_f + (\sigma_M)_{e_{f\ max}} (1 - V_f) \tag{1}$$

where $(\sigma_M)_{e_{f\ max}}$ = matrix stress at a matrix strain equal to the

45

maximum tensile stress in the fibres, so that we can write

$(\sigma_M)_{e_{f\ max}} \equiv \sigma_{M\ y}$, the matrix yield strength

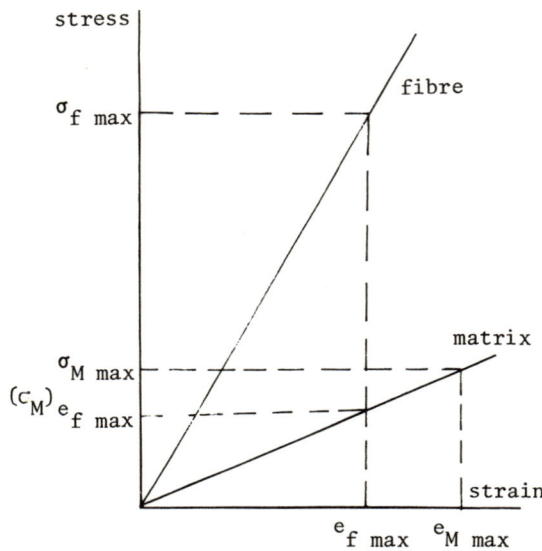

Figure 4.4

If fibre reinforcement is to yield greater strength than can be obtained by the matrix alone, then

$$\sigma_{c\ max} > \sigma_{M\ max} \tag{2}$$

Then equations 1 and 2 can be solved for $V_{f\ crit}$, a critical fibre volume fraction, which must be exceeded to achieve strengthening by fibre reinforcement, giving

$$V_{f\ crit} = \frac{\sigma_{M\ max} - \sigma_{M\ y}}{\sigma_{f\ max} - \sigma_{M\ y}}$$

The composite fails after fracture of the fibres if

$$\sigma_{c\ max} = \sigma_{f\ max} V_f + \sigma_{M\ y} (1 - V_f) \geqslant \sigma_{M\ max} (1 - V_f)$$

from which a minimum value of V_f is obtained as

$$V_{f\ min} = \frac{\sigma_{M\ max} - \sigma_{M\ y}}{\sigma_{f\ max} + \sigma_{M\ max} - \sigma_{M\ y}}$$

PROBLEMS

(1) Indicate with an aid of a sketch how the modulus of elasticity of a composite material loaded at an angle θ to the fibres would be expected to vary for $0° < \theta < 90°$.

(2) Discuss in detail two mechanisms which may account for the high notch toughness of composite materials.

(3) Compare the role of the fibres, particles and matrix in a fibre-reinforced metal and a dispersion particle-hardened metal.

(4) Outline the characteristics of the ideal material for a load-bearing member in a supersonic aircraft. Make special reference to strength/weight, stiffness/weight and temperature stability of various candidate materials.

5   CORROSION

5.1 INTRODUCTION

Consider the case of a metal M in equilibrium with its ions

$$M^{Z+} + Ze \rightleftharpoons M$$

The tendency of a metal to transfer its ions into solution is counterbalanced by an electrode potential difference. Thus equilibrium potentials of metals characterise the free energy change of the process of metal ion removal from metal into solution. In the table below are given some values of normal (standard) equilibrium potentials (that is, where the activity of their own ions in solution is unity) at 298 K; these are known as standard potentials.

Standard electrode potentials of some metals

Metal	Electrode Potential (V)
Magnesium	-2.37 (calculated)
Zinc	-0.77
Iron	-0.44
Copper	0.34
Silver	0.80
Gold	1.50

The relationship of equilibrium potentials to the activity of their metal ions in solution, a_{Z+}, is given by the Nernst equation

$$E = E^0 + 2.3 \frac{RT}{ZF} \log a_{Z+} \qquad (5.1)$$

where E = potential of the metal at an ionic activity equal to $a_M Z+$, E^0 = standard potential of the metal, Z = number of electrons transferred, R, T and F have their usual meanings.

In a more general sense, we can write the Nernst equation as

$$E = E^0 + 2.3 \frac{RT}{ZF} \log \frac{[\text{oxidised state}]}{[\text{reduced state}]}$$

with the square brackets representing concentrations (strictly speaking they should represent activities but for dilute solutions, activity can be assumed equal to concentration). So for the oxygen electrode

$$O_2 + 2H_2O + 4e \rightleftharpoons 4OH^-$$

we can write

$$E = E^0_{O/OH^-} - \frac{2.3RT}{ZF \times 4} \log [a_{OH^-}] + \frac{2.3RT}{ZF \times 4} \log [O_2]$$

5.2 DIFFUSION RATE

When the corrosion is solely dependent on the rate of reduction of dissolved oxygen, the corrosion rate can be calculated from a relationship linking the average diffusion coefficient, D, the number of electrons transferred, Z, the Faraday constant F, the bulk concentration, C, the thickness of the diffusion layer, δ, and the transport number, t.

$$\text{Corrosion rate} = i_{corr} = \frac{DZFC}{(1 - t)\delta} \text{ A m}^{-2}$$

Neglecting migration (t $\approx$ 0)

$$i_{corr} = \frac{DZFC}{\delta} \text{ A m}^{-2} \tag{5.2}$$

5.3 CONCENTRATION OVERPOTENTIAL

Let the deposition of metal ions be carried out at a rate of i A m^{-2}, then i/ZF mol s^{-1} are removed from the double layer. Let the bulk metal ion concentration be C_s and the equilibrium concentration following ion diffusion and migration be $C_s{'}$. Then we can write

$$\text{Deposition rate} = \text{diffusion rate} + \text{migration rate}$$

$$\frac{i}{ZF} = \frac{D(C_s - C_s{'})}{\delta} + \frac{it}{ZF}$$

In the limit, $C_s{'} = 0$ and $i = i_L$ so

$$i_L = \frac{ZFDC_s}{(1 - t)\delta}$$

and $\dfrac{C_s{'}}{C_s} = 1 - \dfrac{i}{i_L}$

From equation 5.1 for concentrations of C_s and $C_s{'}$, we have

$$E = E^0 + \frac{RT}{ZF} \ln C_s$$

$$E{'} = E^0 + \frac{RT}{ZF} \ln C_s{'}$$

then the concentration overpotential, η_{con}, is given by

$$\eta_{con} = E{'} - E = \frac{RT}{ZF} \ln \left(\frac{C_s{'}}{C_s}\right) = \frac{RT}{ZF} \ln \left(1 - \frac{i}{C_L}\right)$$

5.4 POLARISATION

At equilibrium the anodic current density i_a and the cathodic current density i_c are equal and opposite in sense. When equilibrium is disturbed by applying an external voltage, the electrode is said to be polarised. The change of potential across the electrical double layer indicates the extent of this polarisation, and is known as an overpotential, η. The net anodic or cathodic current densities which result from polarising a metal electrode vary as the exponential of the overpotential when η exceeds 30 mV. For $\eta < \pm$ 30 mV, there is a rough linear dependence of i_a and i_c on η, so we can write

$$\eta_a = a + b \ln i_a$$

$$\eta_c = a' - b' \ln i_c$$

where

$$a = (-RT/\beta ZF) \ln i_0$$

$$a' = \left[RT(1 - \beta)ZF \right] \ln i_0 \qquad (5.3)$$

$$b = RT/\beta ZF$$

and $b' = RT/(1 - \beta)ZF$

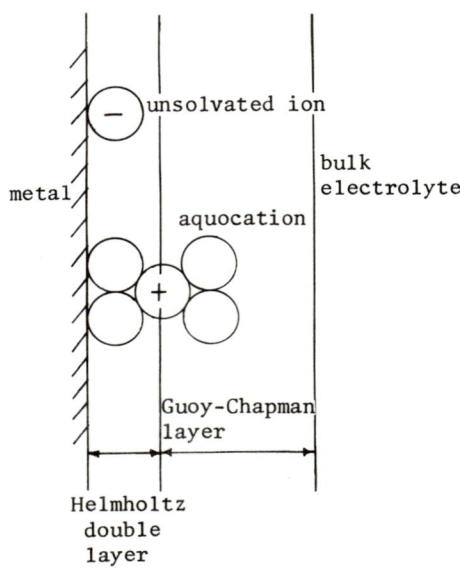

Fig. 5.1

50

These equations are written in the following more familiar form and are then known as Tafel equations

$$\eta_a = b_a \; \log \left(\frac{i_a}{i_0} \right) \tag{5.4}$$

$$\eta_b = b_c \; \log \left(\frac{|i_c|}{i_0} \right) \tag{5.5}$$

with the Tafel constants b_a and b_c being $b_a = 2.3RT/\beta ZF$ and b_c
$b_c = -2.3RT/(1 - \beta)ZF$. In equations 5.3 β represents the fraction of

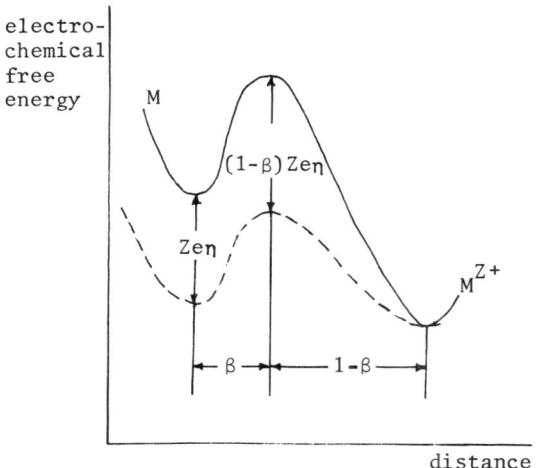

Fig. 5.2

the total distance across the Helmholtz double layer (figure 5.1) where the peak of the energy hump occurs (figure 5.2); it is usually about 0.5. Z is the number of electrons transferred; i_0 is the exchange current density, which is the equilibrium current density when the rate of dissolution is equal to the rate of deposition; R, T and F have their usual meanings.

5.5 THE STERN-GEARY EQUATION

Stern proposed a simple electrochemical method for calculating the corrosion rate of a metal or alloy. It simply involves applying a small potential increment ΔE to a freely corroding specimen and then recording the small current change ΔI accompanying this potential change. The corrosion rate, I_{corr}, expressed as a current, is then given as

$$I_{corr} = \frac{b_a |b_c| \Delta I}{2.3(b_a + |b_c|)\Delta E}$$

Example 5.1

The table below gives the electrode potential in mV for iron immersed in an acid solution of high conductivity and subjected to an impressed current. Given that the polarisation curve is of the form

$$E = a + b \log i$$

(where E = electrode potential, i = current density, a and b are constants) find the corrosion rate, in mg s^{-1}, of a thin iron sheet 100 mm × 100 mm freely corroding in the same solution.

	Current Density, i (mA dm^{-2})		
	1	10	100
Anodic reaction, Fe→Fe^{2+} + 2e	-960	-775	-590
Cathodic reaction, H$_2$↑ on Fe	-370	-640	-910

Atomic weight of iron = 55.8 g mol^{-1}; Faraday constant = 96500 C mol.
(Cambridge)

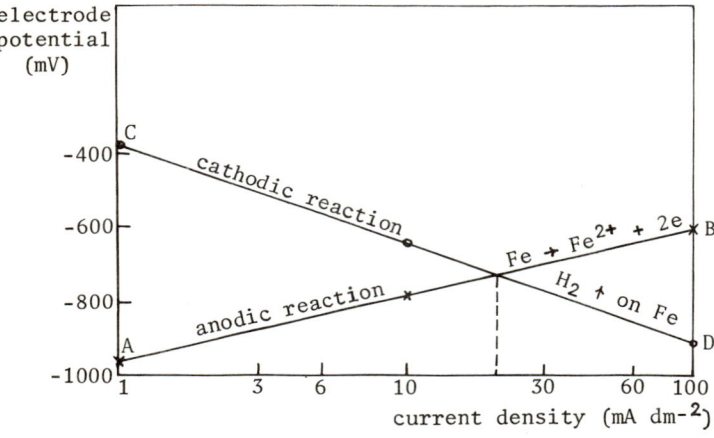

Fig. 5.3

A diagram of E against i is plotted for the two half reactions (figure 5.3) and the intersection on the current density axis is the corrosion current density. In this example it is 20 mA dm^{-2}. Corrosion rate in the required units is therefore

$$\frac{20 \times 10^{-3}}{10^{-2}} \times \frac{55.8 \times 2 \times 10^{-2}}{2 \times 96500} \text{ g s}^{-1} = 11.6 \text{ μg s}^{-1}$$

Example 5.2

The sensitivity of an alloy to compositional variations may be expressed by the parameter

$$S = \frac{\Delta (\log i_{crit})}{\Delta (wt\%Cr)}$$

where i_{crit} is the critical current density. Also S is a function of the pH of the solution in which the alloy is immersed, such that

$$S = A + B \, pH$$

where A and B are constants.

Given that A and B for FeCr alloys in a given solution are -0.12 and -0.04, respectively, calculate the solution pH at which two areas of a chromium steel differing by 5 wt% of chromium exhibit the following corrosion characteristics.

wt%Cr	0	5	10
i_{crit} (mA cm^{-2})	10^3	1	10^{-3}

From the two expressions given for S we have

$$A + BpH = \frac{\Delta (\log i_{crit})}{\Delta (wt\%Cr)}$$

Putting in the values for A, B and $\Delta (\log i_{crit})$ and $\Delta (wt\%Cr)$, we have

$$-0.12 - 0.04 \, pH = -\frac{3}{5}$$

whence solution pH = 12.

Example 5.3

A bimetallic platinum-copper couple is immersed in an acid solution at 25 °C and oxygen is passed into the solution rapidly. Calculate the anodic current density on the copper assuming that the corrosion rate of uncoupled copper in the same solution can be neglected and that the areas of the platinum and copper are 10 cm^2 and 1 cm^2, respectively. The solubility of oxygen in the solution is 1.4 μmole cm^{-3}; the diffusion coefficient of oxygen is 1.75×10^{-5} cm^2 s^{-1}; the thickness of the diffusion layer is 0.05 cm; the Faraday constant is 96500 C per g equivalent.
(London)

Assume that the corrosion rate is determined by the limiting diffusion current density which, for oxygenated solutions, is

$$i_L = \frac{DZFc}{\delta}$$

(neglecting migration) where D = diffusion coefficient of oxygen, Z = number of electrons involved in the electrochemical process, F = Faraday constant, C = concentration of oxygen in solution, δ = thickness of the diffusion layer.

For the dissolution of oxygen in solution, the relevant electro-chemical equation is

$$O_2 + 2H_2O + 4e \rightarrow 4OH^-$$

i.e. $Z = 4$

Thus $i_L = \dfrac{1.75 \times 10^{-5} \times 4 \times 96500 \times 1.4 \times 10^{-6}}{0.05} = 190\ \mu A\ cm^{-2}$

But copper is the anode having an area of $1\ cm^2$, thus

anodic current density $= 190 \times 11 = 2090\ \mu A\ cm^{-2}$

(small anode - large cathode effect)

Example 5.4

Calculate the corrosion rate for a reaction $M + 2H^+ = M^{2+} + H_2$ which is proceeding at 25 °C. The relevant data are as follows.

(a) A change in potential of 10 mV results in a change of current of 5 mA.
(b) Exchange current densities for the hydrogen evolution and metal dissolution reactions are $10^{-10}\ A\ cm^{-2}$ and $10^{-4}\ A\ cm^{-2}$, respectively.
(c) $2.303RT/F = 0.06$ V at 25 °C
(d) Assume that the rate determining step for $H^+ \rightarrow \tfrac{1}{2}H_2\uparrow$ and $M \rightarrow M^{2+}$ involves one and two electrons, respectively and that the transfer coefficient α is 0.5 for both half reactions constituting the corrosion reaction.
(London)

The Stern-Geary principle of evaluating corrosion rate of a metal is given by the equation

$$\frac{\Delta E}{\Delta I_{applied}} = \frac{b_a|b_c|}{2.3 I_{corr}(b_a + |b_c|)}$$

Now $\alpha = 0.5$ for both the anodic and cathodic reactions so $b_a = 0.06$ V per decade and $b_c = -0.12$ V per decade. Applying these data to the Stern-Geary equation, we have

$$2 = \frac{0.06 \times 0.12}{2.3 \times 0.18 I_{corr}}$$

whence $I_{corr} = 8.7$ mA.

Example 5.5

Given that the penetration rate is 0.637 s μm^{-1} at 25 °C, calculate the corrosion current density developed during the measurement of copper thickness in a jet test at this temperature. Density = 8.95 g cm^{-3}; Faraday constant = 96500 C mole^{-1}; copper molar mass = 63.5 g.

$$\text{Dissolving copper loss rate} = \frac{63.5i}{96\ 500 \times 2} \text{ g cm}^{-2} \text{ s}^{-1}$$

where i = corrosion current density, A cm^{-2}.

$$\text{Copper penetration rate} = \frac{63.5i}{96\ 500 \times 8.95 \times 2} \text{ cm s}^{-1}$$

Thus $\dfrac{10^{-4}}{0.637} = \dfrac{63.5i}{96\ 500 \times 8.95 \times 2}$

whence i = 4.27 A cm^{-2}

Example 5.6

A mild steel surface is completely immersed in aerated seawater which flows over and parallel to the surface at a steady rate. It is to be assumed that under these conditions the untreated surface will provide roughly equal cathodic and anodic areas.

It is to be further assumed that the worst conditions for corrosion will obtain when dissolution of the steel as ferrous ions occurs at a rate equivalent to the current density afforded by the complete reduction of dissolved oxygen. Data appropriate to the system are given below.

Mean seawater properties

ρ, density	1100 kg m^{-3}
μ, viscosity	0.0012 N s m^{-2}
D, diffusivity of oxygen	2×10^{-9} m^2 s^{-1}
mean oxygen content	0.5×10^{-3} kion m^{-3}
d, equivalent diameter of surface	0.10 m
V, mean velocity of seawater	0.30 m s^{-1}
Faraday's number	96 MC (kequiv)$^{-1}$

The analogy for mass transport at place surfaces is given as

$$\frac{K}{V}\left(\frac{\mu}{\rho D}\right)^{2/3} = 0.005 \left(\frac{Vd\rho}{\mu}\right)^{-0.20}$$

where K = mass transfer coefficient for dissolved oxygen.

(a) Determine the corrosion rate, in suitable units, at the anodic surfaces. (b) Indicate the probable effect on the corrosion rate of a five-fold increase in seawater velocity.

(a) Inserting numerical values for the parameters in the equation of the analogy for mass transport, we have

$$\frac{K}{0.3} \times \left(\frac{0.0012}{1100 \times 2 \times 10^{-9}}\right)^{2/3} = 0.005 \left(\frac{0.3 \times 0.1 \times 1100}{0.0012}\right)^{-0.2}$$

whence $K = 0.29 \times 10^{-5}$ m s^{-1}

Corrosion rate $= 0.29 \times 10^{-5} \times 0.5 \times 10^{-3} \times 96 \times 10^{6} = 140$ mA m^{-2}

(b) The analogy for mass transport at plane surface can be written as

$$K = \text{constant} \times V^{-0.2} \times V$$

or $\quad K = \text{constant} \times V^{0.8}$

So $\quad \dfrac{K_2}{K_1} = \left(\dfrac{V_2}{V_1}\right)^{0.8}$

For a five-fold increase in seawater velocity

$$K_2 = K_1(5)^{0.8} = 3.62K_1$$

For the same mean oxygen content, the corrosion rate is directly proportional to K so that for a five-fold increase in seawater velocity the corrosion rate is increased by about 262%.

Example 5.7

A corrosion cell contains a steel sheet measuring 75 mm × 25 mm × 0.1 mm thick immersed in dilute acid. The corrosion current is measured as 210 μA. Calculate how long it will be before the metal dissolves in the corrosive medium. Density of steel = 7.87 g cm^{-3}; Faraday constant = 96500 C mol^{-1}; Atomic weight of iron = 55.8 g mol^{-1}.

$$\text{Weight of dissolving metal} = \frac{55.8}{2} \times \frac{210 \times 10^{-6}}{96500} \times t \ \text{g} \qquad (1)$$

If metal is completely dissolved then

$$\text{weight of dissolved metal} = 7.87 \times 7.5 \times 2.5 \times 0.01 \ \text{g} \qquad (2)$$

From equations 1 and 2

$$\frac{55.8 \times 210 \times 10^{-6}t}{2 \times 96\,500} = 7.87 \times 7.5 \times 2.5 \times 0.01$$

$$t = 24 \times 10^{6} \ \text{s}$$

whence required time = 278 days.

Example 5.8

If the corrosion potential of an alloy when immersed in seawater (which is approximately a 3% $NaCl$ solution) is 0.6 V vs SHE, calculate whether it will corrode in this medium if the chloride breakdown potential is given as

$$E_b = 1.2 - n \log (Cl^-)$$

where $n = 2$ and (Cl^-) is the chloride concentration of the solution, expressed in moles per litre. Molecular weight of $NaCl = 58.5$ g mole-

3% $NaCl \equiv 30$ g l^{-1} $NaCl$. Molecular weight of $NaCl = 58.5$ g mole^{-1}, so 30 g l^{-1} $NaCl = 0.51$ M l^{-1}. Chloride breakdown potential, E_b, is now

$$E_b = 1.2 - 2 \log (0.51) = 1.78 \text{ V}$$

The alloy will not corrode in this medium because the corrosion potential, 0.6 V, is less than the breakdown potential of 1.78 V.

Example 5.9

It is possible to carry out certain polymerisation reactions both electrochemically and thermally. In one such case the rate of the activation-controlled electrochemical process is increased by a change in the overpotential of +0.5 V. Calculate the corresponding temperature increase, from room temperature of 25 °C, required to bring about the same change in the reaction rate. The Tafel gradient can be taken as 60 mV per decade and the activation energy as 80 kJ mol^{-1}; the molar gas constant, $R = 8.314$ J mol^{-1} K^{-1}. (Nottingham)

The anodic overpotential is given by

$$\eta = b_a \log \left(\frac{i}{i_0}\right)$$

Applying this to the two cases, we have

$$\Delta\eta = b_a \log \left(\frac{i_2}{i_1}\right)$$

i.e. $0.5 = 0.06 \log \left(\dfrac{i_2}{i_1}\right)$

thus $\log \left(\dfrac{i_2}{i_1}\right) = 8.33$

Assuming that the corrosion rate follows an Arrhenius law, we can write

$$i = A \exp \left(-\frac{Q}{RT}\right)$$

57

thus $\log \left(\dfrac{i_2}{i_1}\right) = \dfrac{Q}{2.3R} \left(\dfrac{1}{T_1} - \dfrac{1}{T_2}\right)$

$8.33 = 4.2 \times 10^3 \left(\dfrac{1}{298} - \dfrac{1}{T_2}\right)$

whence $T_2 = 739$ K. Hence accompanying temperature increase is 441 °C.

Example 5.10

A steel tank is to contain concentrated sulphuric acid in a manufacturing operation. Under simulated conditions in the laboratory the corrosion current of the steel has been measured to be 8.68 μA cm^{-2}. Calculate the effective rate of corrosion of the steel in mm year^{-1}.

Equivalent weight of iron = 27.9 g mol^{-1}

Dissolving metal loss rate = $\dfrac{8.68 \times 10^{-6} \times 27.9}{96\ 500}$

$= 2.5 \times 10^{-9}$ g cm^{-2} s^{-1}

Rate of corrosion of steel = $\dfrac{2.5 \times 10^{-9} \times 3600 \times 24 \times 365 \times 10}{7.87}$

$= 0.1$ mm year^{-1}

Example 5.11

Given that a particular paint film has a porosity of 0.001%, calculate the impressed current required to protect 300 m^2 of steel piling immersed in seawater. Assume that 0.5 V are required for protection and that the Tafel slope is 0.06 V. The exchange current density of the anode is 10^{-8} A m^{-2}.
(Nottingham)

The anodic overpotential is

$\eta = b_a \log \left(\dfrac{i}{i_0}\right)$

With $b_a = 0.06$ V and $i_0 = 10^{-8}$ A m^{-2}, we have

$0.5 = 0.06 \log \left(\dfrac{i}{10^{-8}}\right)$

whence $i = 2.14$ A m^{-2}.

Exposed area = $\dfrac{0.001}{100} \times 300 = 3 \times 10^{-3}$ m^2

thus required current for protection = $3 \times 10^{-3} \times 2.14 = 6.42$ m A

Example 5.12

The diffusion rate of a metal anode dissolving in an electroplating solution is 10^2 A m^{-2} at an overpotential of 0.75 V. Assuming that the overpotential is entirely due to activation energy barriers, determine the increase in current density when the overpotential is increased from 0.75 V to 1 V. Take F = 96 500 C mol^{-1}; R = 8.314 J mol^{-1} K^{-1}; T = 298 K; β = 0.5; Z = 1.

First calculate the Tafel slope as

$$b_a = \frac{2.3RT}{\beta ZF} = \frac{2.3 \times 8.314 \times 298}{0.5 \times 96\ 500} = 120 \text{ mV per decade}$$

Applying the Tafel equation

$$\eta = b_a \log \left(\frac{i}{i_0}\right)$$

to the two cases, we have

$$\Delta\eta = b_a \log \left(\frac{i_2}{i_1}\right)$$

$$0.25 = 0.12 \log \left(\frac{i_2}{i_1}\right)$$

$$i_2 = 121 i_1$$

Example 5.13

A rotating cylinder is used as a cathode in an electrolyte bath under conditions of turbulent flow. Assuming that the cathodic deposition reaction is under diffusion control, the following relationship obtains

$$\frac{i_L}{ZFC_b U} = f \left(\frac{rU\rho}{\mu}, \frac{\mu}{\rho D}\right)$$

where i_L = limiting current density, Z = number of electrons involved, F = Faraday's constant, r = radius of the cylinder, C_b = bulk concentration, U = peripheral velocity, ρ = electrolyte density, D = diffusion coefficient, μ = electrolyte viscosity.

Under conditions of constant Reynolds number, the relationship reduces to

$$\frac{i_L}{ZFC_b U} = k \left(\frac{\mu}{\rho D}\right)^{-0.6}$$

where k is a constant.

When copper was electrodeposited on to a rotating cylinder the following data were obtained

Limiting current density	600 A m^{-2}
Kinematic viscosity (μ/ρ)	1.2 × 10^{-6} m^2 s^{-1}
Diffusion coefficient for copper ions	5.2 × 10^{-10} m^2 s^{-1}
Peripheral velocity	0.3 m s^{-1}
Concentration of CuSO$_4$	70 mole m^{-3}

Calculate the mass transfer coefficient between a rotating iron cylinder dissolving in carbon-saturated molten iron, if the Reynolds number in this and the above experiments were the same, using the data below

Diffusion coefficient for iron ions	9 × 10^{-9} m^2 s^{-1}
Kinematic viscosity	1.4 × 10^{-6} m^2 s^{-1}
Peripheral velocity	0.18 m s^{-1}
Faraday constant	96 500 C mole^{-1}

(Nottingham)

Under conditions of constant of Reynolds number, we have

$$\frac{i_L}{ZFC_b U} = k \left(\frac{\mu}{\rho D}\right)^{-0.6}$$

Applying this to the electrodeposition involving copper and iron, we can write

$$\frac{600}{2 \times 96\ 500 \times 70 \times 0.3} = k \left(\frac{1.2 \times 10^{-6}}{5.2 \times 10^{-10}}\right)^{-0.6} \tag{1}$$

$$\text{and} \quad \frac{i_L}{2 \times 96\ 500 \times C_b \times 0.18} = k \left(\frac{1.4 \times 10^{-6}}{9 \times 10^{-9}}\right)^{-0.6} \tag{2}$$

From equations 1 and 2

$$\frac{i_L}{C_b} = 26 \text{ A m mole}^{-1}$$

The required mass transfer coefficient for the deposition of iron is given by

$$\frac{i_L}{C_b ZF} = \frac{26}{2 \times 96\ 500} \text{ m s}^{-1} = 13.5 \times 10^{-5} \text{ m s}^{-1}$$

Example 5.14

A tube of a metal which exhibits an active-passive transition is used for transferring an oxygenated solution from one vessel to another.

During operation the solution passes through the tube at a very high velocity, but during shut-down periods the solution is static. Comment on the suitability of this metal for the required use. Limiting current density for passivation = 75 μA cm^{-2}; thickness of diffusion layer = 0.05 cm when static, and 0.005 cm when the solution is flowing at high velocity; solubility of oxygen in the solution = 0.5 μmole cm^{-3}; diffusion coefficient of oxygen = 1.5 × 10^{-5} cm^2 s^{-1}; 1 Faraday = 96500 C (g equivalent)$^{-1}$. (London)

The relevant reaction in oxygenated water is

$$O_2 + 2H_2O + 4e \rightarrow 4OH^-$$

and the corrosion rate will be equal to i_L, the limiting current density for oxygen reduction. Now $i_L = DZFC/\delta$, thus in static conditions

$$i_{L_s} = \frac{1.5 \times 10^{-5} \times 4 \times 96\ 500 \times 0.5 \times 10^{-6}}{0.05} = 57.9\ \mu\text{A cm}^{-2}$$

In flowing conditions

$$i_{L_f} = \frac{1.5 \times 10^{-5} \times 4 \times 96\ 500 \times 0.5 \times 10^{-6}}{0.005} = 579\ \mu\text{A cm}^{-2}$$

But i_L for passivation is 75 μA cm^{-2}. During shut-down periods, the corrosion density is less than i_L, the passivation current density, while during flow operation it is greater than i_L; hence the metal is suitable, since, during operation, i_L is exceeded and the metal will passivate.

Example 5.15

Calculate the diffusion overpotential for silver depositing at a rate of 7 × 10^{-2} g dm^{-2} min^{-1} from a cyanide solution at 25 °C. Assume that the current efficiency is 100% and that the limiting current density is 2 A dm^{-2}. (1 coulomb liberates 0.001 118 g of silver and 2.3RT/F at 25 °C = 0.060 V.)

The diffusion overpotential is given by

$$\eta_{diff} = \frac{2.3RT}{ZF} \log \left(1 - \frac{i}{i_L}\right) \tag{1}$$

Now $i = \dfrac{7 \times 10^{-2}\ \text{g A s}}{\text{dm}^2 \times 60\ \text{s} \times 0.001\ 118\ \text{g}} = 104\ \text{Am}^{-2}$

and $i_L = 200$ A m^{-2}

Putting these values in equation 1, we have

$$\eta_{diff} = 60 \log \left(1 - \frac{104}{200} \right) = -19 \text{ mV}$$

Example 5.16

Calculate the corrosion rate in A cm^{-2} for the following reaction, which proceeds isothermally at 25 $^\circ$C

$$M + 2H^+ = M^{2+} + H_2$$

Data: (a) For $M \rightleftharpoons M^{2+}$, $i_0 = 10^{-3}$ A cm^{-2}, $\alpha = 0.5$; (b) for $H^+ + e \rightleftharpoons H_2$, $i_0 = 10^{-10}$ A cm^{-2}, $\alpha = 0.5$; (c) 2.3 RT/F = 0.060 V at 25 $^\circ$C.

For the anodic process the Tafel equation is

$$\eta_a = a + b \log i$$

For the cathodic process it is

$$\eta_c = a - b \log i$$

Thus, we have

$$\eta_a = \left(\frac{0.06}{0.5 \times 2} \times (-3) \right) + \frac{0.06}{0.5 \times 2} \log i$$

$$\eta_a = -0.18 + 0.06 \log i \text{ V} \qquad (1)$$

$$\eta_c = \left(\frac{0.06}{0.5 \times 1} \times (-10) \right) - \frac{0.06}{0.5 \times 1} \log i$$

$$\eta_c = -1.2 - 0.12 \log i \text{ V} \qquad (2)$$

At corrosion point, $\eta_a = \eta_c$, thus

$$-0.18 + 0.06 \log i = -1.2 - 0.12 \log i$$

whence corrosion rate i = $10^{-5.67}$ A cm^{-2}.

Example 5.17

The polarisation of a copper electrode in dilute unstirred cupric sulphate is caused by charge transfer and diffusion process. For current density, i, greater than 10^{-4} A cm^{-2}, the charge transfer overpotential is

$$\eta_{ct} = a \pm b \log |i| \text{ V}$$

where a and b are constants. A limiting current density is observed for copper deposition, $i_L = -10^{-5}$ A cm^{-2}, and the diffusion over-potential is given by

$$\eta_{diff} = \pm\, 0.03\ \log\left(1 - \frac{i}{i_L}\right)\ V$$

Show that over a range of current densities, the total anodic polarisation is

$$\eta_{total} = a' + b'\ \log\,|i|$$

and hence evaluate the values of a' and b'. (Note that over-potentials and current densities follow the convention, +ve anodic, -ve cathodic.)
(London)

The total anodic polarisation, $\eta_{total} = \eta_{ct} + \eta_{diff}$

$$\eta_{ct} = a + b\ \log\ i$$

$$\eta_{diff} = 0.03\ \log\left(1 - \frac{i}{i_L}\right)$$

$$= -0.03\ \log\ i + 0.03\ \log\ i_L$$

Now $i_L = -10^{-5}$ A cm^{-2}, hence

$$\eta_{diff} = -0.03\ \log\ i + 0.15$$

Then $\eta_{total} = a + b\ \log\ i - 0.03\ \log\ i + 0.15$

$$= (a + 0.15) + (b - 0.03)\log\ i$$

$$= a' + b'\ \log\ i$$

where $a' = (a + 0.15)$V, $b' = (b - 0.03)$V.

PROBLEMS

(1) Outline the factors that are important for electrodeposition of metals from aqueous solutions.

Copper has a molecular weight of 63.54 g mole^{-1} and a density of 8.96 g cm^{-3}. What thickness of copper is deposited in 10 min at a current density of 100 mA cm^{-2}? Faraday constant = 96 500 C mole^{-1}.
[0.0022 cm]

(2) Calculate the concentration of dissolved oxygen in mole dm^{-3} that would be required to passivate a stainless steel specimen with

a critical current density in a static solution, at 25 °C, of 0.5 A m^{-2}, and indicate qualitatively how this would be affected by solution temperature and agitation. Diffusion coefficient of oxygen at 25 °C = 10^{-9} m2 s^{-1}; thickness of diffusion layer in static solution = 0.5 mm; Faraday constant = 96 500 C mole^{-1}.
(Nottingham)   [0.65 × 10^3 mole dm^{-3}]

(3) The linear polarisation method is being used to estimate the corrosion rate of a metal of area 0.01 m^2 in an acid solution at 25 °C. In the pure acid solution it is found that when the potential of the metal is changed by 5 mV the current produced is 10 mA, but that this falls to 1 mA when an inhibitor is added to the solution. Calculate the corrosion rates of the metals (in A m^{-2}) in the two solutions. For hydrogen evolution at the surface of the metal the transfer coefficient, $\alpha = 1/2$ and the number of electrons involved $z = 1$; for the anodic dissolution of the metal $\alpha = 1/2$ and $z = 2$. RT $\ln x/F = 0.06 \log x$ at 25 °C.
(Nottingham)   [3.48 A m^{-2}; 0.35 A m^{-2}]

(4) During the testing of a steel structure the following results were obtained

Anode potential (mV)	5	7	10	12
Current in external circuit (A)	38.5	53.9	78.1	97.0

It was assumed that the Tafel slopes for both anode and cathode reactions were 60 mV per decade. Calculate the corrosion current.

Outline the limitations and assumptions in the Stern-Geary method of obtaining corrosion currents.
(Nottingham)   [109 A]

64

6   OXIDATION

6.1 INTRODUCTION

Oxidation is the chemical reaction of gases, such as oxygen, sulphur, nitrogen, with a metal or alloy surface to form a surface film or oxide. Oxides formed at low temperatures are generally thin and those formed at high temperatures are generally thick. One of the main features of oxidation is that the oxides are formed directly on those parts of the metal or alloy surface that enter into the reaction. Further growth of the oxide film will depend on the ability of the oxidant to penetrate the film. One condition that limits the ability of the initial film to resist further oxidation of the underlying surface is the continuity of the film. In this respect, Pilling and Bedworth gave the empirical rule that if

$$\frac{\text{volume of oxide, } V_{0x}}{\text{volume of metal forming the oxide, } V_m} > 1$$

the film is continuous, protective and further oxidation is retarded. If, however, $V_{0x}/V_M < 1$, the film is porous, discontinuous and un-protective. Some values of this ratio are as follows.

Metal	Oxide	Pilling-Bedworth Ratio
Aℓ	Aℓ$_2$O$_3$	1.28
Cu	Cu$_2$O	1.64
W	WO$_3$	3.35*
K	K$_2$O	0.45
Mg	MgO	0.81

*Too high a ratio leads to brittle oxide films and continuity can be lost as a result of high internal stresses.)

6.2 OXIDATION KINETICS

The linear law of film growth states that for metals that do not form protective oxides, the rate of growth of the oxide remains constant, i.e.

$$\frac{dy}{dt} = K_1$$

where y = film thickness, t = time of oxidation, and K_1 is a constant. Integrating

$$y = K_1 t + A$$

where A is a constant of integration. An example of this type of law is magnesium oxidation in oxygen at 503 °C.

The parabolic law of film growth states that where protective films are formed, the oxidation process is resisted by the diffusion of the oxidant through the film, resulting in film thickening at a rate which continuously decreases with time of oxidation. Thus

$$\frac{dy}{dt} = \frac{K_2}{y}$$

or $y^2 = 2K_2t + K_3$

where K_2 is a material constant and K_3 is a constant of integration. An example of this type of law is iron oxidation in air at 1100 °C.

Film thickness is proportional to the oxide weight, Δw, so that the linear and parabolic laws can be written

$$\Delta w = K_1 t + A$$

and $(\Delta w)^2 = 2K_2t + K_3$

Other rate laws include the cubic: $y^3 = K_3 t$ and logarithmic: $y = K_4 \ln(K_5 t + K_6)$ where K_3, K_4, K_5 and K_6 are constants.

6.3 OXIDATION THERMODYNAMICS

Consider the oxidation reaction

$$M + O \rightarrow MO$$

If the reaction potential is E_0 we can write the relationship

$$\Delta G = -ZFE_0$$

where ΔG = Gibbs free energy of the reaction. For a spontaneous oxidation E_0 will be positive since ΔG is negative.

The van't Hoff isotherm expresses the relationship between the Gibbs energy and the concentrations of the reactants and products as

$$\Delta G = \Delta G^0 + RT \ln \frac{[\text{products}]}{[\text{reactants}]}$$

where ΔG^0 = value of ΔG when all the effective concentrations, denoted by the square brackets, are unity.

Example 6.1

Magnesium parts are heat-treated in an environment containing 5% sulphur dioxide, forming a thin sulphate coating. Show why this procedure is preferable to heating in air. Densities of $MgSO_4$, MgO and Mg, in g cm^{-3}, are 2.66, 3.58 and 1.74, respectively. Atomic weights of $MgSO_4$, MgO and Mg, in g $mole^{-1}$, are 120.39, 40.32 and 24.32, respectively.

66

The approach is to calculate the Pilling-Bedworth ratio for $Mg \rightarrow MgSO_4$ and $Mg \rightarrow MgO$.

For $Mg \rightarrow MgSO_4$, the Pilling-Bedworth ratio is

$$\frac{V_{MgSO_4}}{V_{Mg}} = \frac{120.39 \times 1.74}{2.66 \times 24.32} = 3.24$$

For $Mg \rightarrow MgO$, the Pilling-Bedworth ratio is

$$\frac{V_{MgO}}{V_{Mg}} = \frac{40.32 \times 1.74}{3.58 \times 24.32} = 0.81$$

In the latter case, the Pilling-Bedworth ratio is less than unity, so MgO would be unprotective, while MgSO4 would be protective because the Pilling-Bedworth ratio is greater than unity. However, the value of 3.24 is too large so there might well be a tendency for MgSO4 to crack; although, because the atmosphere is not 100%, this possibility is small.

Example 6.2

Show by calculation whether the nitrification of titanium might be expected to be linear or parabolic (in terms of its oxidation). The relevant chemical reaction is

$$2Ti + N_2 \rightarrow 2TiN$$

The densities of titanium and titanium nitride are 4.50 g cm^{-3} and 5.43 g cm^{-3}, respectively. The atomic weights of titanium and titanium nitride are 47.90 g mol^{-1} and 61.91 g mol^{-1}, respectively.

The Pilling-Bedworth ratio for $Ti \rightarrow TiN$ is

$$\frac{V_{TiN}}{V_{Ti}} = \frac{61.91 \times 4.50}{47.90 \times 5.43} = 1.07$$

This ratio is slightly greater than unity so that probably an adherent scale is formed. The nitrification would thus probably be parabolic.

Example 6.3

The following crystallographic data refer to two metals and their oxides.

Material	Structure	No. of Metal Atoms per Unit Cell	Unit Cell Dimension($\overset{\circ}{A}$)	
			a	c
Ca	f.c.c.	-	5.582	--
Cd	h.c.p.	-	2.979	5.618
CaO	cubic	4	4.780	-
CdO	cubic	4	4.680	-

Using the above data and on the basis of the Pilling-Bedworth rule comment on the protective ability of the two oxides.

Cadmium Cadmium metal is h.c.p. so there are 2 atoms per unit cell. The volume of the unit cell is $\frac{1}{2}\sqrt{3}[a(Cd)^2c(Cd)]\overset{\circ}{A}^3$. Thus the atomic volume is $\frac{1}{4}\sqrt{3}[a(Cd)]^2[c(Cd)]\overset{\circ}{A}^3$.

Cadmium oxide is cubic and there are 4 atoms per unit cell. The volume of the unit cell is $[a(CdO)]^3\overset{\circ}{A}^3$ so that the atomic volume is $\frac{1}{4}[a(CdO)]^3\overset{\circ}{A}^3$. Thus Pilling-Bedworth ratio is

$$\frac{V_{CdO}}{V_{cd}} = \frac{(4.68)^3}{\sqrt{3} \times (2.979)^2 \times 5.618} = 1.18$$

Calcium Calcium metal is f.c.c. so there are 4 atoms per unit cell. Volume of unit cell is $[a(Ca)]^3\overset{\circ}{A}^3$ so atomic volume is $\frac{1}{4}[a(Ca)]^3\overset{\circ}{A}^3$.

Calcium oxide is cubic and there are 4 atoms per unit cell. Volume of unit cell is $[a(CaO)]^3\overset{\circ}{A}^3$ and atomic volume is $\frac{1}{4}[a(CaO)]^3\overset{\circ}{A}^3$. Thus Pilling-Bedworth ratio is

$$\frac{V_{CaO}}{V_{Ca}} = \frac{(4.78)^3}{(5.582)^3} = 0.63$$

The Pilling-Bedworth ratio is greater than unity in the case of Cd→CdO but less than unity for Ca→CaO. Thus CdO is likely to be protective while CaO is likely to be unprotective.

Example 6.4

The oxidation rates of zirconium diboride, ZrB_2, and zirconium carbide, ZrC, at elevated temperatures have been studied by recording the increase in weight of these materials in pure oxygen as a function of time. Using the experimental data presented below, suggest possible mechanisms of oxidation for these materials.

Time (min)	Weight Increase (mg cm^{-2})	
	ZrB_2 (at 945 °C)	ZrC (at 652 °C)
10	1.01	0.444
40	2.02	1.776
80	2.85	3.551

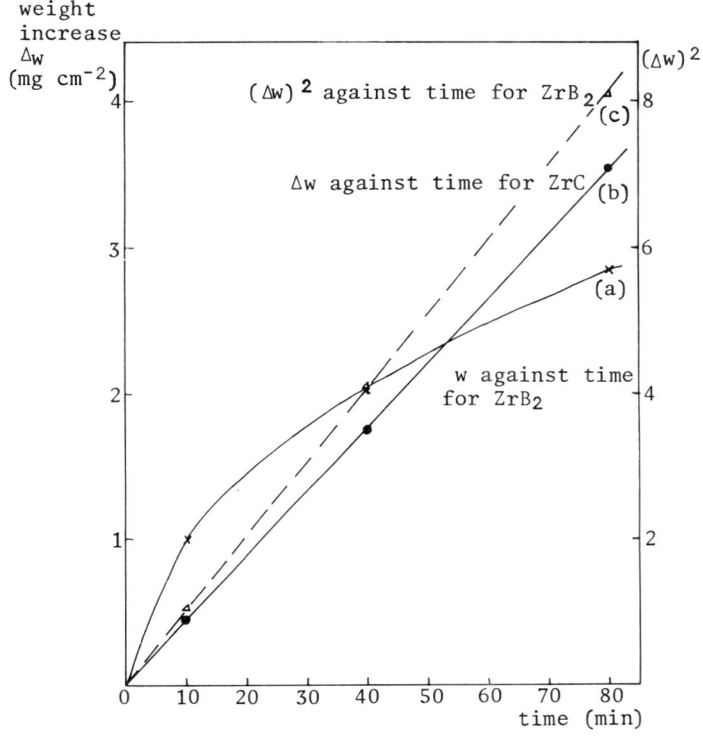

Figure 6.1

A graph of Δw (weight increase) against time, t, is plotted for each of the two materials (figure 6.1). This gives a parabola for ZrB_2 (line (a)) and a straight line for ZrC (line (b)). This indicates that the oxidation laws are, respectively, parabolic and linear. To further test the applicability of the parabolic law, a graph of $(\Delta w)^2$ against t for ZrB_2 is plotted. The relevant data points are

$(\Delta w)^2$ $(mg^2\ cm^{-4})$	t (min)
1.02	10
4.08	40
8.12	80

This graph (line (c)) is a straight line, thus confirming that the oxidation of ZrB_2 at 945 °C is indeed parabolic.

Suggested mechanisms: (a) ZrB_2 - parabolic oxidation, (b) ZrC - linear oxidation.

Example 6.5

In the temperature range 300 K to 1500 K a metal M forms two oxides MO and M_2O_5. The oxidation rate constants for the growth of the oxides were found to vary with the temperature and pressure according to the following expressions

$$k(MO) = 10^3 a_{O_2}^{1/6} \exp (-240\ 000/RT)\ kg^2\ m^{-4}\ s^{-1}$$

$$k(M_2O_5) = 10^{-5} a_{O_2}^{-1/6} \exp (-120\ 000/RT)\ kg^2\ m^{-4}\ s^{-1}$$

where k = rate constant (in terms of the mass of oxide formed), a_{O_2} = activity of oxygen.

From this and the data given below, calculate the relative thicknesses of the two oxides obtained at 1000 K and under a pressure of 0.1 atm of oxygen.

density of M_2O_5 = 5.0 kg dm^{-3}

density of MO = 8.0 kg dm^{-3}

molar gas constant, R = 8.314 J mol^{-1} K^{-1}

Let $a_{O_2} = P_{O_2}$, the pressure of oxygen. Then at 1000 K and P_{O_2} = 0.1 atm

$$k(M_2O_5) = 10^{-5}(0.1)^{-1/6} \exp \left(\frac{-120}{8.314}\right) = 7.91 \times 10^{-12}\ kg^2\ m^{-4}\ s^{-1}$$

and $$k(MO) = 10^3 (0.1)^{1/6} \exp \left(\frac{-240}{8.314}\right) = 1.98 \times 10^{-10}\ kg^2\ m^{-4}\ s^{-1}$$

The oxidation is parabolic, thus we can write $(\Delta w)^2 = kt$ which, in terms of density ρ and thickness ℓ becomes

$$\frac{\rho^2 (M_2O_5)\ell^2 (M_2O_5)}{\rho^2 (MO)\ell^2 (MO)} = \frac{k(M_2O_5)}{k(MO)}$$

(For the same cross-sectional area and time, t)
Thus

$$\frac{\ell (M_2O_5)}{\ell (MO)} = \sqrt{\left(\frac{7.91 \times 10^{-12} \times 64}{1.98 \times 10^{-10} \times 25}\right)} = 0.32$$

Example 6.6

Calculate the life of a strip of metal used as a heating element in a furnace, given the following information. The oxidation rate constant of the metal (given in terms of the weight of oxygen) was found to vary with temperature according to the expression

$$k = 10^6 \exp(-260\ 000/RT)\ kg^2\ m^{-4}\ s^{-1}$$

The relevant data are: weight fraction of oxygen in the oxide = 0.2; density of the oxide = 6.0 kg dm^{-3}; Pilling-Bedworth ratio = 1.5; gas constant, R = 8.314 J mole^{-1} K^{-1}; furnace temperature = 1027 °C; volume of metal strip = 3 × 10^{-4} m^3.

Volume of metal oxide, V_{MO} = 1.5 × volume of metal

$$= 4.5 \times 10^{-4} \text{ m}^3$$

mass of metal oxide per unit area, $\Delta m_{MO} = \rho_{MO} V_{MO}$

where ρ_{MO} is density of metal oxide. Therefore

$$\Delta m_{MO} = 6.0 \times 10^3 \times 4.5 \times 10^{-4} = 2.7 \text{ kg m}^{-2}$$

Weight fraction of oxygen in oxide $= \dfrac{\text{mass of oxygen in oxide}}{\text{mass of metal oxide}}$

therefore

mass of oxygen in oxide Δm_0 = 0.2 × 2.7 = 0.54 kg m^{-2}

From the given rate constant, the oxidation law is parabolic. That is

$$(\Delta m)^2 = Kt + C$$

where Δm = weight, K = rate constant, t = time, C is a constant of integration.

The oxidation of the metal is governed by the change in weight of oxygen. Thus we can write the parabolic oxidation law as

$$(\Delta m_0)^2 = Kt + C$$

At t = 0, Δm_0 = 0, therefore C = 0 and

$$(\Delta m_0)^2 = Kt$$

thus $(0.54)^2 = 10^6 \exp\left(\dfrac{-260\ 000}{8.314 \times 1300}\right) \times t$

whence life of metal strip, t = 2.27 h.

Example 6.7

An atmosphere used for the heat treatment of mild steel is composed of N_2, CO_2, H_2O and H_2, where the CO/CO_2 and H_2O/H_2 ratios are 0.2 and 0.5, respectively. Determine the temperature region in which the metal can be annealed in the atmosphere without the risk of oxidation. State all assumptions made. The relevant data are as follows

$$Fe(s) + \tfrac{1}{2}O_2(g) = FeO(s), \quad \Delta G_1 = -270 + 0.065T \text{ kJ} \tag{1}$$

$$H_2(g) + \tfrac{1}{2}O_2(g) = H_2O(g), \quad \Delta G_2 = -250 + 0.055T \text{ kJ} \tag{2}$$

$$CO(g) + \tfrac{1}{2}O_2(g) = CO_2(g), \quad \Delta G_3 = -280 + 0.080T \text{ kJ} \tag{3}$$

The molar gas constant $R = 8.314$ J mole^{-1} K^{-1}.

From equations 1 and 2, we have

$$Fe + H_2O = FeO + H_2, \quad \Delta G_4{}^0 = -20 + 0.010T \text{ kJ}$$

Now $\Delta G_4 = \Delta G_4^o + RT \ln \left(\dfrac{a_{FeO}\, a_{H_2}}{a_{Fe}\, a_{H_2O}} \right)$

so that, with a_{FeO} and a_{Fe} each equal to 1, we have

$$\Delta G_4 = -20 + 0.010T + (8.314 \times 2.3T \log 2)10^{-3}$$

$$= -20 + 0.0158T \text{ kJ} \tag{4}$$

From equations 1 and 3, we have

$$Fe + CO_2 = FeO + CO$$

and $\quad \Delta G_5 = 10 - 0.015T + 8.314 \times 2.3T \times 10^{-3} \log 5$

$$= 10 - 0.002T \text{ kJ} \tag{5}$$

There must be a temperature where the oxidation tendency as given by equation 4 is equal to the reduction tendency as given by $-\Delta G_5$ in equation 5. Thus

$$-20 + 0.0158T = -10 + 0.002T$$

$$T = 725 \text{ K}$$

The required temperature range is $725 < T <$ melting point of iron.

Example 6.8

An alumina crucible is to be used in the vacuum annealing of a pure niobium strip. From that data given below, calculate

(a) the maximum temperature at which the heat treatment could be safely performed without the oxidation of the niobium, and
(b) the equilibrium pressure of oxygen at this temperature.

72

$$\tfrac{4}{3}A\ell + O_2 = \tfrac{2}{3}A\ell_2O_3, \quad \Delta G_1 = -1118 + 0.245T \text{ kJ} \tag{1}$$

$$2Nb + O_2 = 2NbO, \quad \Delta G_2 = -841 + 0.18T \text{ kJ} \tag{2}$$

(a) Maximum temperature is given by the intersection of the lines describing the $\Delta G^0/T$ relationships, i.e.

$$-1118 + 0.245T = -841 + 0.18T$$

or $T = 4262$ K

(b) To calculate the equilibrium pressure, p_{O_2} we use equation 2

$$-841 + 0.18T = -RT\,\ell n\ K$$

$$(-841 + 0.18 \times 4262)10^3 = -2.3 \times 8.314 \times 4262\ \log\left(\frac{1}{p_{O_2}}\right)$$

whence $p_{O_2} = 0.12$ atm.

Example 6.9

Given that

$$PdO\ (solid) \rightleftharpoons Pd\ (solid) + \tfrac{1}{2}O_2\ (gas)$$

and $\Delta G^0 = 114.3 - 0.10T$ kJ

calculate (a) the dissociation pressure as a function of T; (b) the decomposition temperature, T_1, at $P_{O_2} = 1$ atm; (c) the decomposition temperature, T_2, at $P_{O_2} = 0.21$ atm.

(a) $\Delta G^0 = 114.3 - 0.10T = -RT\,\ell n\left(\dfrac{a_{Pd}a_{O_2}^{\frac{1}{2}}}{a_{PdO}}\right) = -RT\ \ell n\ a_{O_2}^{\frac{1}{2}}$

$(a_{Pd} = a_{PdO} = 1)$

$$\Delta G^0 = -8.314 \times 2.3 \times T \log p_{O_2}^{\frac{1}{2}} = -19.12T \log P_{O_2}^{\frac{1}{2}}$$

So $114\ 300 - 100T = -19.12T \log p_{O_2}^{\frac{1}{2}}$

and $P_{O_2} = 10^{-(11956/T - 10.46)} \tag{1}$

(b) Substitute $p_{O_2} = 1$ atm in equation 1 to give

$$\frac{-11956}{T_1} + 10.46 = 0$$

73

or $T_1 = 1143$ K

(c) Substitute $p_{O_2} = 0.21$ atm in equation 1 to give

$$\log (0.21) = -\frac{11956}{T_2} + 10.46$$

or $T_2 = 1073$ K

Example 6.10

Given the reaction

$$\tfrac{4}{3}MO + O_2 \rightleftharpoons \tfrac{2}{3}M_2O_5$$

$$\Delta G^0 = -500 + 0.1T \text{ kJ}$$

Calculate (a) pressure of oxygen and hence (b) chemical potential of oxygen at 1000 K.

(a) At 1000 K, $\Delta G^0 = -400$ kJ. But $\Delta G^0 = -RT \ln K$, thus

$$-400 \times 1000 = -2.3 \times 8.314 \times 1000 \log K$$

thus $K = 10^{20}$

Now $K = \dfrac{a_{M_2O_5}^{2/3}}{a_{MO}^{4/3} \times a_{O_2}}$

Assume pure M_2O_5 and MO and putting $a_{O_2} = P_{O_2}$ we have

$$P_{O_2} = 10^{-20} \text{ atm}$$

(b) Chemical potential, $\Delta\mu$, is obtained from

$$\mu_i = \mu_i^0 + RT \ln a_{O_2}$$

$$\Delta\mu = 2.3 \times 8.314 \times 1000 \log 10^{-20} = -400 \text{ kJ}$$

Example 6.11

In a solid solution of metals A and B, containing 0.3 mole fraction of A the activity coefficients of A and B are 0.375 and 0.825, respectively. If the alloy is maintained in an atmosphere of H_2O/H_2 vapour (steam), determine (a) which of the components is more likely to oxidise at 1000 K, and (b) the minimum p_{H_2}/p_{H_2O} ratio that must be maintained to avoid the oxidation of the alloy.

74

$$2A(s) + O_2(g) \rightleftharpoons 2AO, \quad \Delta G_1 = -800 + 0.26T \text{ kJ} \tag{1}$$

$$2B(s) + O_2(g) \rightleftharpoons 2BO, \quad \Delta G_2 = -500 + 0.18T \text{ kJ} \tag{2}$$

$$2H_2(g) + O_2(g) \rightleftharpoons 2H_2O, \quad \Delta G_3 = -500 + 0.12T \text{ kJ} \tag{3}$$

(a) The objective here is to establish which component has the higher equilibrium constant since it is that one which will oxidise first. From equations 1 and 3, we have

$$A + H_2O \rightarrow AO + H_2, \quad \Delta G^0 = -150 + 0.07T \text{ kJ} \tag{4}$$

At 1000 K, $\Delta G^0 = -80$ kJ. Also

$$\Delta G^0 = -2.3 \times 8.314 \times 1000 \times \log k_p$$

whence $k_p = 10^4$.

From equations 2 and 3, we have

$$B + H_2O \rightarrow BO + H_2, \quad \Delta G^0 = 0.03T \text{ kJ} \tag{5}$$

From equation 4, we conclude straight away that oxidation is not possible (because ΔG^0 is positive). The answer to part (a) is thus that A will oxidise and B will dissociate.

(b) Knowing that $k_p = 10^4$ we can calculate the minimum value of the ratio p_{H_2}/p_{H_2O} by putting

$$10^4 = \frac{a_{AO} \times a_{H_2}}{a_A \times a_{H_2O}}$$

Now activity, $a = \gamma x$ (γ = activity coefficient, x = mole fraction). Thus $a_A = 0.1125$. Assume $a_{AO} = 1$, $a_{H_2} = p_{H_2}$ and $a_{H_2O} = p_{H_2O}$, then

$$\frac{p_{H_2}}{p_{H_2O}} = 10^4 \times 0.1125 = 1125$$

PROBLEMS

(1) Figure 6.2 shows an idealised weight gain against time curve for a metal oxidising at constant temperature. For each section of the curve describe a model to account for the observed kinetics.

What factors lead to the changes from one form of kinetics to another?
(Nottingham)

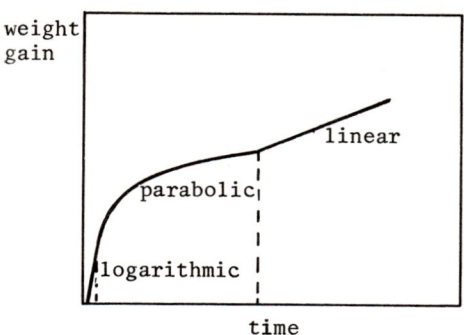

Figure 6.2

(2) Write a short essay on the oxidation resistance of metals, ceramics and polymers.

(3) The oxidation of a certain alloy follows the law

$$x^2 = kt$$

where x = oxide thickness, t = time, k = $\exp[-15500/T(K)]$ cm^2 s^{-1}. For an oxidation temperature, T, of 650 °C calculate how long it will take to grow a film 0.1 cm thick on the surface.

[54.6 h]

7   CREEP AND CREEP STRESS RELAXATION

7.1 INTRODUCTION

The time-dependent increase in strain at constant stress is known as
creep, whereas the decrease in stress at constant strain is known as
creep stress relaxation. Both of these phenomena are very pronounced
at high temperatures and can operate continuously over very long
periods.

7.2 CREEP

Andrade was the first to propose a relationship between creep strain
and time in order to describe the primary and secondary creep stages.
This is given as

$$\text{creep strain, } \varepsilon = \alpha t^{1/3} + \beta t \tag{7.1}$$

where α and β are material constants, t = time. Graham and White
extended the analysis to cover the tertiary stage with the relationship

$$\varepsilon = \alpha t^{1/3} + \beta t + \gamma t^3 \tag{7.2}$$

where γ is another material constant; α, β and γ are functions of
stress and temperature.

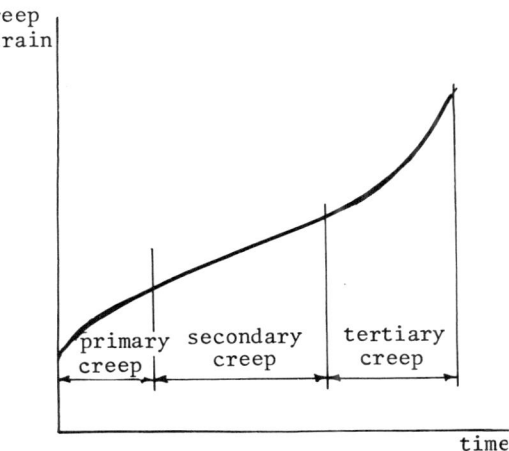

Figure 7.1

A number of theories have been proposed to explain the secondary creep stage but the rate-process theory is the best known. This suggests that secondary creep strain is a thermally activated process. Thus we can write an Arrhenius equation

$$\text{creep strain rate, } \dot{\varepsilon} = A \exp\left(-\frac{Q}{RT}\right) \tag{7.3}$$

where A is a material constant, Q = activation energy for the creep process, R and T have their usual meanings. Larsen and Miller have analysed the above relationship at constant stress, σ, to give

$$\left(\frac{Q}{R}\right)_\sigma = T(\ln A + \ln t - \ln \varepsilon)$$

$$\left(\frac{Q}{R}\right)_\sigma = T(B + \ln t) \tag{7.4}$$

for a given value of strain. The right-hand side of equation 7.4 is known as the Larsen-Miller parameter and plotting $\ln \sigma$ against this parameter yields straight lines for different strains. Such a plot is known as a master creep curve.

At constant temperature, the empirical relationship between secondary creep strain rate and applied stress is

$$\dot{\varepsilon} = K\sigma^n \tag{7.5}$$

where K and n are material constants.

7.3 CREEP STRESS RELAXATION

Let initial stress be σ_0 and initial strain be e_0. Assume initial elastic strain in the structure, constant temperature and no deformation. After some time, t, has elapsed the stress in the structure will fall to the current stress σ. The strain is constant. The effect of creep, therefore, is to induce a plastic strain, e, which is the strain remaining after removing the recoverable elastic strain, σ/E. Thus the compatibility equation for strain becomes

$$e_0 = e + \left(\frac{\sigma}{E}\right)$$

Differentiating with respect to time gives

$$\frac{de_0}{dt} = \frac{de}{dt} + \frac{1}{E}\frac{d\sigma}{dt}$$

Strain is constant, so we have

$$0 = \dot{e} + \frac{1}{E}\frac{d\sigma}{dt}$$

78

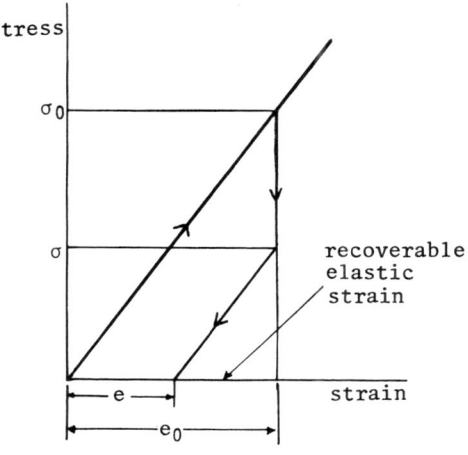

Figure 7.2

For creep in the secondary stage (from equation 7.5)

$$\dot{e} = K\sigma^n$$

Then $EK\sigma^n \, dt = -dt$

thus $$\int_0^T dt = -\int_{\sigma_0}^{\sigma} \frac{d\sigma}{EK\sigma^n}$$

$$T = \frac{1}{EK(n-1)}\left(\frac{1}{\sigma^{n-1}} - \frac{1}{\sigma_0^{n-1}}\right)$$

Example 7.1

A lead wire 7.5 mm diameter is used as a fuse for a delayed-action grenade (figure 7.3). The wire is in series with a steel spring having a spring constant of 25 N mm⁻¹. Before the fuse is set, point B is held so that there is no stress in the wire. When the safety pin is removed, spring BC immediately exerts an axial force of 450 N on the wire. When the spring relaxes 2.5 mm, the grenade explodes.

Calculate the delay period of the grenade. Assume that secondary creep of lead can be expressed as

$$\dot{e} = A\left(\frac{S}{S_0}\right)^n \quad mm \ mm^{-1} \ min^{-1}$$

where $A = 2.68 \times 10^{-5} \ min^{-1}$, $S_0 = 7 \ MN \ m^{-2}$ and $n = 7.68$.

79

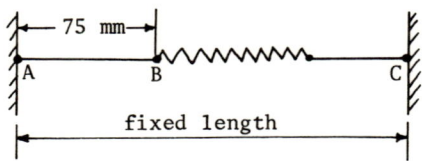

Figure 7.3

Force exerted when wire relaxes x mm = 450 -25 xN

Strain, $e = \dfrac{x}{75}$

$de = \dfrac{dx}{75}$

But $e = A\left(\dfrac{s}{SO}\right)^n$

Thus $S_0{}^n \dfrac{de}{dt} = AS^n$

$S_0{}^n \dfrac{dx}{75} = AS^n \, dt$

Now $S = \dfrac{450 - 25x}{\text{area}}$

so $\dfrac{(S_0 \times \text{area})^n}{75A} \, 2{\cdot}5 \int_0^T dx = \int_0^T (450 - 25x)^n \, dt$

Where T is the grenade delay period.   Thus

$T = \dfrac{1.34 \times 10^{19}}{6.68 \times 25 \times 75 \times 2.68 \times 10^{-5}} \left(\dfrac{1}{(387.5)^{6\cdot68}} - \dfrac{1}{(450)^{6\cdot68}}\right)$

= 130 min

Example 7.2

The secondary stage creep behaviour of a 12% steel at 500 °C is shown
(approximately) in figure 7.4. Investigate whether there is a
simple power-law relationship between the creep rate and the stress.
(Cambridge)

A calculation of the creep strain rates at various stresses from
figure 7.4 gives

Stress, σ (N mm^{-2})	Creep Strain Rate, $\dot{\varepsilon}$ (min^{-1})
175	0.025×10^{-5}
205	0.055×10^{-5}
235	0.100×10^{-5}
255	0.200×10^{-5}

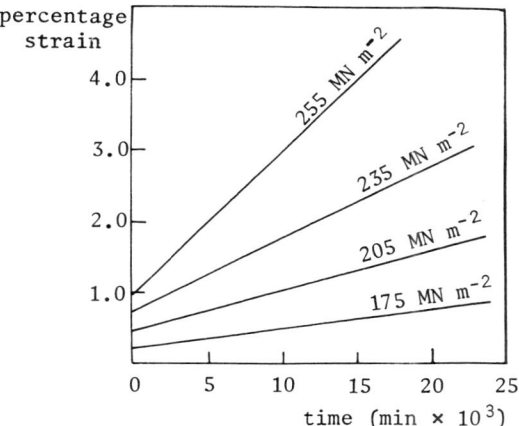

percentage strain

time (min × 10^3)

Figure 7.4

The relationship between stress and creep strain rate is

$$\dot{\epsilon} = B\sigma^n$$

We now have

$\log \dot{\epsilon}$	-6.60	-6.26	-6.00	-5.70
$\log \sigma$	7.24	2.31	2.37	2.41

Example 7.3

Given the stress-deformation data for a low-carbon austenitic alloy forged disc at 816 °C (figure 7.5), calculate the minimum creep rate at a stress of 105 MN m^{-2}.

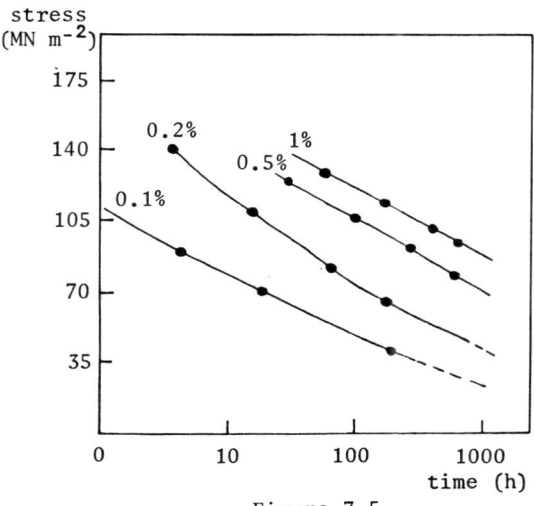

stress (MN m^{-2})

time (h)

Figure 7.5

Assuming a straight line to these data points, we have

$$\dot{\varepsilon} = 10^{-18.14} \sigma^{5.13} \text{ min}^{-1}$$

At 816 °C the following values are read from the curves in figure 7.5 at a stress of 105 MN m^{-2}.

Time (h)	Elongation (%)
1.5	0.1
20.0	0.2
110.0	0.5
230.0	1.0

Then from a plot of % elongation against time (figure 7.6) we obtain the gradient of the linear portion as 0.0033% h^{-1}, thus minimum creep rate at 10 MN m^{-2} = 33 × 10^{-4} % h^{-1}.

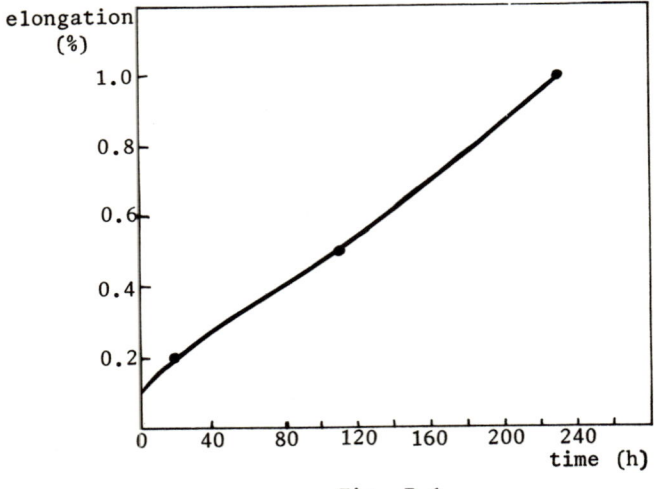

Fig. 7.6

Example 7.4

A thin-walled polymer pipe is subjected to a steady internal pressure of 700 kN m^{-2} at 20 °C. If a tensile stress of 17.5 MN m^{-2} is not to be exceeded and the internal radius is 100 mm, what will be the increase in diameter after 1000 h? The tensile creep curves for the polymer provide the following values at 1000 h (neglecting creep contraction).

Stress (MN m^{-2})	6.9	13.8	20.7	27.6	34.5
Strain (%)	0.20	0.48	0.97	1.72	3.38

The tensile creep values are plotted in figure 7.7. For a thin-walled tube, tensile stress is given as pD/2t, where p = internal pressure, D = diameter, t = thickness. In the case of a tensile stress of 17.5 MN m^{-2}, we have

$$17.5 = \frac{0.7 \times 200}{2t}$$

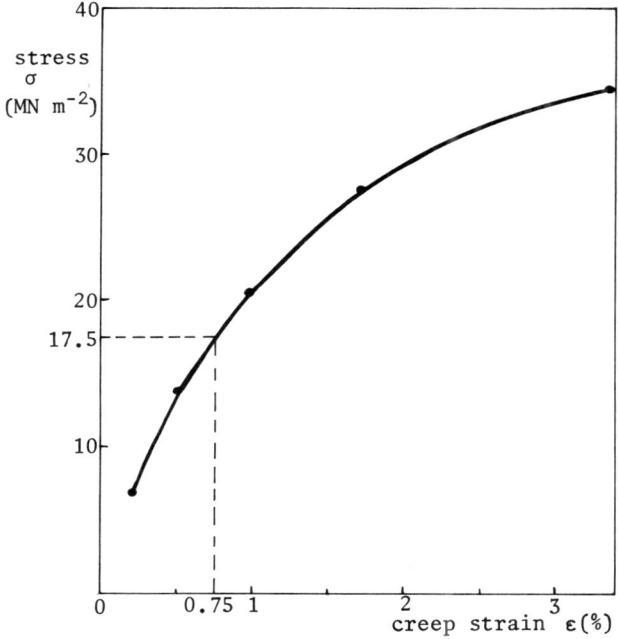

Fig. 7.7

giving a thickness of 4 mm, so that the original diameter is 208 mm.

From the tensile creep graph (figure 7.7) at 17.5 MN m^{-2}, the creep strain is 0.75%. Thus we have

$$\frac{0.75}{100} = \frac{\text{increase in diameter}}{208}$$

whence increase in diameter = 1.56 mm.

Example 7.5

A series of creep tests at constant stress on an Aℓ-Cu alloy gave the following results for the minimum creep strain rate.

83

Temp (°C)	140	180	220	260	300
Minimum creep strain rate (h^{-1})	0.04	0.17	0.75	3.26	6.57

Suggest a suitable analytical form for the relationship between creep rate and temperature and evaluate all the constants in such a relationship. (R, molar gas constant = 8.314 J mol^{-1} K^{-1}.)

A suitable analytical form is

$$\dot{\varepsilon} = A \exp \left(-\frac{Q}{RT} \right)$$

where $\dot{\varepsilon}$ = creep strain rate, Q = activation energy (J mol^{-1}), R = molar gas constant (J mol^{-1} K^{-1}), T = absolute temperature (K), A is a material constant. The equation can then be written as

$$\log \dot{\varepsilon} = \log A - \frac{Q}{2.3RT} \tag{1}$$

Thus a plot of $\log \dot{\varepsilon}$ against $1/T$ would yield both log A and Q.

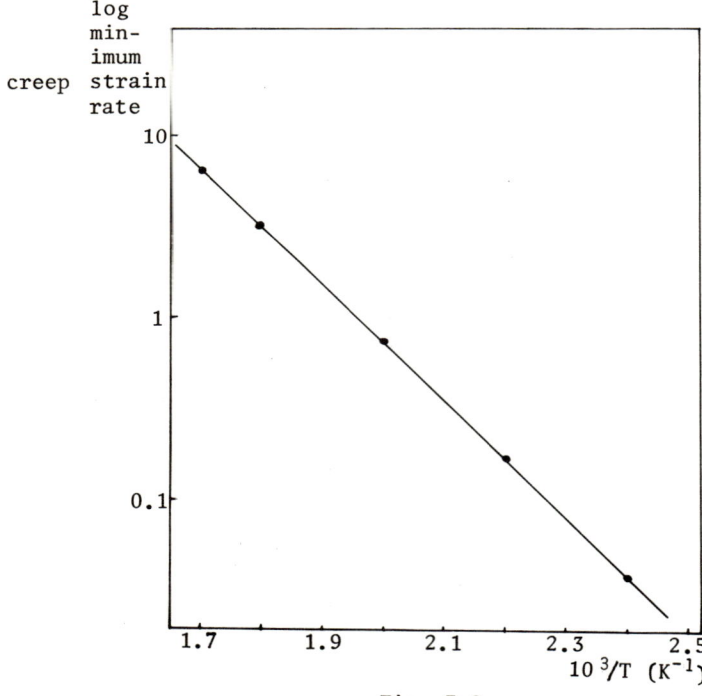

Fig. 7.8

84

The test results can now be put in a suitable form for plotting.

$\dot{\varepsilon}$ (h^{-1}) 0.04   0.17   0.75   3.26   6.57
1/T (K^{-1}) 0.0024 0.0022 0.0020 0.0018 0.0017

A graph is then plotted (figure 7.8) from which

$$-\frac{Q}{2.3R} = -\text{slope}$$

$$Q = 2.3 \times 8.314 \times 3.18 \times 10^3$$

$$= 61 \text{ kJ mol}^{-1}$$

A is now obtained by inserting the value of E into equation 1 for any corresonding values of $\dot{\varepsilon}$ and T, thus A = $10^{6.35}$ h^{-1}.

Example 7.6

Figure 7.9 presents data on the creep of an aluminium alloy at 176 °C for two stress levels.

(a) Calculate the minimum creep strain rate for a stress of 59 MN m^{-2}.
(b) A Newtonian fluid has a constant value of viscosity at a constant temperature. Where the viscosity decreases as the stress increases, the flow is non-Newtonian of the pseudoplastic type. In the case where the viscosity increases with increase in stress, the flow is non-Newtonian of the dilatant type.

Given that the minimum creep strain rate $\dot{\varepsilon}$ is related to the stress, σ and the viscosity η by the expression $\dot{\varepsilon} = \sigma/3\eta$, determine the character of the flow during the steady-state period of the creep of the alloy.

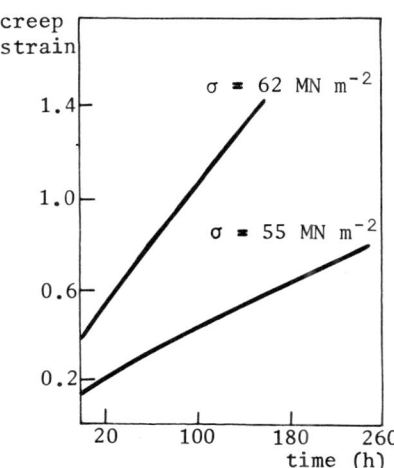

Figure 7.9

85

(a) The minimum creep strain rate, $\dot{\varepsilon}$, of the material at any given stress, σ, at constant temperature is given by

$$\dot{\varepsilon} = B\sigma^n \qquad (1)$$

Consider the given curves at 62 MN m^{-2} and 55 MN m^{-2} (figure 7.9). At 62 MN m^{-2}, $\dot{\varepsilon} = 0.0066$ h^{-1} and at 55 MN m^{-2} $\dot{\varepsilon} = 0.0025$ h^{-1}. Inserting these values in equation 1, we have

$$0.0066 = B \times 62^n$$

$$0.0025 = B \times 55^n$$

giving $n = 8.1$ and $B = 2 \times 10^{-17}$ h^{-1}. At 59 MN m^{-2}, the minimum creep strain rate is

$$2 \times 10^{-17}(59)^{8.1} = 0.0044 \text{ h}^{-1}$$

(b) $\dot{\varepsilon} = \sigma/3\eta$ and $\eta = \sigma/3\dot{\varepsilon}$, then

At 55 MN m^{-2}, $\eta = \dfrac{55 \times 10^6 \times 3600}{3 \times 0.0025} = 26.4 \times 10^{12}$ N s m^{-2}

At 59 MN m^{-2}, $\eta = \dfrac{59 \times 10^6 \times 3600}{3 \times 0.0044} = 16.1 \times 10^{12}$ N s m^{-2}

At 62 MN m^{-2}, $\quad = \dfrac{62 \times 10^6 \times 3600}{3 \times 0.0066} = 11.3 \times 10^{12}$ N s m^{-2}

Viscosity decreases as stress increases, thus the flow is non-Newtonian of the pseudoplastic type.

Example 7.7

The following creep data were obtained in tests carried out at 750 °C on an austenitic steel.

Stress (MN m^{-2})	Minimum Creep Rate (% h^{-1})
70	8×10^{-5}
100	26×10^{-4}
140	25×10^{-3}
205	2
275	3
345	320

A tie-rod is made from this steel and in service at 750 °C is subject to an axial load of 35 kN. Using a factor of safety of 3 find the cross-sectional area of the tie-rod based on the criterion of an allowable creep of 1% in 10 000 h.

The creep strain rate-stress relationship is

$$\dot{\varepsilon} = B\sigma^n$$

Hence we can plot log $\dot{\varepsilon}$ against log σ (figure 7.10). The creep strain rate, $\dot{\varepsilon} = 10^{-4}$ % h^{-1} so that log $\dot{\varepsilon}$ = -4. Log σ at this creep strain rate is 1.88 (figure 7.5) so the stress at this strain rate is 75.86 MN m^{-2}.

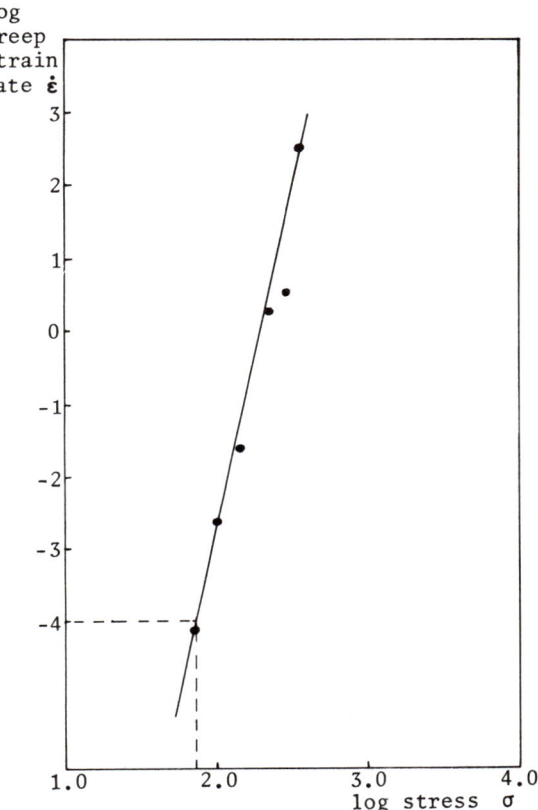

Fig. 7.10

The factor of safety is 3 so that we can write

$$\text{working stress} = \frac{\text{working load}}{\text{cross-sectional area}}$$

i.e. $75.86 = \dfrac{105\ 000\ \text{N}}{\text{Area mm}^2}$

whence cross-sectional area = 1400 mm^2.

Example 7.8

From the body of data given below determine the maximum operating temperature such that failure by creep of the alloy should not occur

87

in 5000 h at stress levels of 200 MN m^{-2}.

Temperature (°C)	Stress (MN m^{-2})	Rupture Time (h)
650	480	22
650	480	40
650	480	65
650	450	75
650	380	210
650	345	2700
650	310	3500
705	310	275
705	310	190
705	240	960
705	205	2050
760	205	180
760	205	450
760	170	730
760	140	2150
815	140	29
815	140	45
815	140	65
815	120	90
815	120	115
815	105	260
815	105	360
815	105	1000
815	105	700
815	85	2500
870	83	37
870	83	55
870	69	140
870	42	3200
980	21	440
1095	10	155

We use the Larsen-Miller creep rupture parameter $T(20 + \log t)$, where $T(K)$ is the operating temperature and $t(h)$ the rupture time. This parameter is calculated for each of the experimental temperature-stress-rupture time data points. The results are plotted (figure 7.11) as log stress against the parameter.

At a stress level of 200 MN m^{-2} the Larsen-Miller parameter is 23000. Thus

$$T(20 + \log 5000) = 23\ 000$$

whence maximum operating temperature = 698 °C.

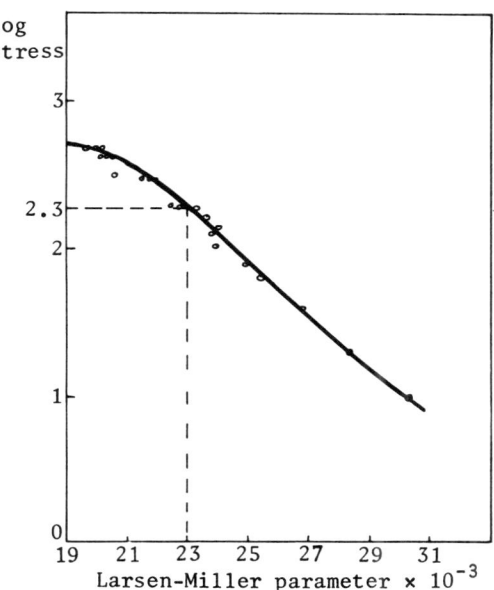

$$\begin{array}{c} \text{log} \\ \text{stress} \end{array}$$

Fig. 7.11

Example 7.9

A pressure vessel accommodates a chemical process at a pressure of 1.03 MN m^{-2} and temperature 455 °C. One of the dished ends of the vessel has a 40 cm diameter man-hole placed at its centre and a cover plate is held in position by twenty chrome-molybdenum steel bolts, 2.5 cm diameter, equally spaced around the flanges, which may be regarded as being relatively rigid.

Calculate from first principles the initial tightening stress in the bolt so that after 10 000 h of creep stress relaxation there is still a factor of safety of 2 against leakage. Also determine the time for leakage to occur.

Neglect primary creep and assume that secondary creep for the bolt steel at 455 °C may be represented by

$$\dot{e} = B\sigma^n$$

where $B = 4.5 \times 10^{-15}$ (MN m^{-2})$^{-n}$ h^{-1}, $\dot{e}$ = creep rate (h^{-1}), σ = stress (MN m^{-2}), $n = 4$. Take Young's modulus, E, for the steel to be 172 GN m^{-2}.

For compatibility of forces

force in man-hole = force in bolts

$$1.03 \times \frac{\pi}{4} \times 40^2 = 20\sigma_{\text{bolts}} \times \frac{\pi}{4} \times 2.5^2$$

whence σ_{bolts} = 13.2 MN m^{-2}.

With a factor of safety of 2 the current tightening stress is now 26.4 MN m^{-2}. Thus, for current strain

$$e_0 = e + \frac{\sigma}{E}$$

where e_0 = initial strain, σ = initial stress.

$$\frac{de_0}{dt} = \dot{e} + \frac{1}{E}\frac{d\sigma}{dt}$$

$$0 = \dot{e} + \frac{1}{E}\frac{d\sigma}{dt}$$

thus $B\sigma^n = -\frac{1}{E}\frac{d\sigma}{dt}$

Stress falls from σ_0 to σ in time t so that

$$t = \frac{1}{BE(n-1)}\left(\frac{1}{\sigma^{n-1}} - \frac{1}{\sigma_0^{n-1}}\right)$$

$$10\ 000 = \frac{1}{4.5 \times 10^{-15} \times 172 \times 10^3 \times 3}\left(\frac{1}{26.4^3} - \frac{1}{\sigma_0^3}\right)$$

whence σ_0 = 31.77 MN m^{-2}

For leakage to occur, the stress now falls from 31.77 MN m^{-2} to 13.2 MN m^{-2}, i.e.

$$t = \frac{1}{BE(n-1)}\left(\frac{1}{\sigma^{n-1}} - \frac{1}{\sigma^{n-1}}\right)$$

With the numerical values of the various parameters

$$t = \frac{1}{4.5 \times 10^{-15} \times 172 \times 10^3 \times 3}\left(\frac{1}{13.2^3} - \frac{1}{31.77^3}\right)$$

$$= 173\ 816\ h$$

i.e. time for leakage to occur is 173 816 h.

Example 7.10

A gasket of non-Hookean material of thickness, ℓ, is inserted between the flanges of a pipeline joint. Steel bolts, each of free length L, are used to hold the flanges together. The total cross-sectional area of the bolts is A_s and the total area of gasket in contact with the flanges is A_g. It may be assumed that the stresses

90

in the bolts and gasket are uniform and that the flanges are incompressible.

Show that the creep strain rate $\dot{e}_g$ for the gasket is related to the current compressive stress rate $\dot{\sigma}_g$ by the equation

$$\dot{e}_g = - \left(\frac{LA_g}{E\ell A_s}\right)\dot{\sigma}_g$$

where E is the Young's modulus for steel.

Given that ℓ = 4 mm, L = 75 mm, A_g/A_s = 16 and E = 200 GN m^{-2}, estimate the time for the stress in the gasket to fall by 25% from an initial value of 4.4 MN m^{-2}. For the gasket material the creep rate is $\dot{e}_g = 7 \times 10^{-11}[\sigma_g/\text{MN m}^{-2}]^{7\cdot5}$ h^{-1}.
(London)

Two equations are necessary for the solution of the problem, namely, compatibility of displacement and force.

(a) Compatibility of displacement: for the steel bolts

$$\text{displacement} = \frac{\sigma_s}{E} \times L$$

For the gasket

$$\text{displacement} = e_g \times \ell$$

(since the gasket material is non-Hookean). Therefore for equilibrium

$$\frac{\sigma_s}{E} L = e_g \ell \tag{1}$$

(b) Compatibility of force

$$\text{force in steel bolts} = A_s\sigma_s$$

$$\text{force in gasket} = A_g\sigma_g$$

Thus for equilibrium

$$A_s\sigma_s + A_g\sigma_g = 0 \tag{2}$$

From equation 2

$$\sigma_s = - \frac{A_g}{A_s}\sigma_g$$

Substituting this in equation 1 gives

$$- \frac{A_g \sigma_g L}{EA_s} = e_{\ell} g$$

$$e_g = - \left(\frac{A_g \sigma_g L}{A_s E \ell} \right)$$

Differentiating this, we get

$$\dot{e}_g = - \left(\frac{A_g L}{A_s \ell E} \right) \dot{\sigma}_g$$

Using the given figures for the various quantities, we obtain

$$- \frac{(16 \times 75 \times 10^{-3})}{4 \times 10^{-3} \times 200 \times 10^3} \frac{d\sigma_g}{dt} = 7 \times 10^{-11} \times \sigma_g^{7 \cdot 5}$$

σ_g varies from 4.4 MN m^{-2} to 3.3 MN m^{-2} thus

$$t = \frac{0.214 \times 10^8}{6.5} \times 0.3575 \times 10^{-3} = 1170 \text{ h}$$

Example 7.11

A tension member is to be made of a nylon plastic. It is to be 1 m long and 1290 mm^2 in cross-section. Calculate the allowable load if the total elongation should not exceed 6.94 mm in 1 year.

The creep-stress-time relation for the material is approximately defined by the equation

$$e = \frac{S}{E} + k_1 S^m + k_2 t S^n$$

where $k_1 = 34.8 \times 10^{-9}$, $m = 1.16$, $k_2 = 3.41 \times 10^{-9}$, $n = 0.75$, $E = 1.7$ GN m^{-2}, $S = $ stress. The units of k_1 and k_2 and t are in m and days, respectively.

The creep-stress-time relationship is

$$e = \frac{S}{E} + k_1 S^m + k_2 t S^n$$

Putting in the values of the various parameters, we have

$$\frac{6.94}{10^3} = \frac{S}{17 \times 10^8} + 34.8 \times 10^{-9} S^{1 \cdot 16} + 3.41 \times 10^{-9} \times 365 S^{0 \cdot 75}$$

$$6.94 \times 10^{-3} - 5.88 \times 10^{-10} S = 34.8 \times 10^{-9} S^{1 \cdot 16}$$

$$+ 1.24 \times 10^{-6} S^{0 \cdot 75}$$

92

A numerical solution of the above equation gives S = 25.5 kN m^{-2}, then

$$\text{allowable load} = 25.5 \times 10^3 \times 1290 \times 10^{-6} \text{ N} = 33 \text{ N}$$

Example 7.12

Andrade's relationship

$$\varepsilon = \varepsilon_0 + \alpha t^{1/3} + \beta t$$

can be used to represent strain during creep. In a creep test in which ε_0 was 0.002 11, the strain was 0.002 82 at t = 10 s and 0.005 76 at t = 1000 s. Determine the creep strain rate at t = 100 s. (London)

Putting in the numerical values the Andrade relationship becomes

$$0.00282 = 0.00211 + \alpha \times 10^{1/3} + 10\beta \tag{1}$$

$$0.00576 = 0.00211 + \alpha \times 1000^{1/3} + 1000\beta \tag{2}$$

whence $\alpha = 3.28 \times 10^{-4}$ and $\beta = 3.7 \times 10^{-7}$ s^{-1}.

Differentiating Andrade's equation, we have

$$\dot{\varepsilon} = \frac{1}{3}\alpha t^{-2/3} + \beta$$

so that at t = 100 s

$$\dot{\varepsilon} = \frac{1}{3} \times 3.28 \times 10^{-4} \times 100^{-2/3} + 3.7 \times 10^{-7}$$

$$= 54.4 \times 10^{-7} \text{ s}^{-1}$$

Example 7.13

The secondary creep strain-rate, $\dot{\varepsilon}$, of a metal subjected to a tensile stress, σ, may be expressed as

$$\dot{\varepsilon} = K\sigma^n$$

where K and n are material constants at a given temperature. Determine an expression for the time required for the stress to relax at constant strain from σ_0 to σ_t, where   is the initial stress and σ_t is the stress at time t after the application of stress. (Cambridge)

Initial stress = σ_0

initial elastic strain = e_0

For constant strain

$$e_0 = e + \left(\frac{\sigma}{E}\right)$$

where e = plastic strain and E = Young's modulus. Differentiating with respect to time

$$\frac{de_0}{dt} = \frac{de}{dt} + \frac{1}{E}\frac{d\sigma}{dt}$$

i.e. $0 = \dot{e} + \frac{1}{E}\frac{d\sigma}{dt}$

$$\dot{e} = -\frac{1}{E}\frac{d\sigma}{dt}$$

For creep in the secondary region

$$K\sigma^n = -\frac{1}{E}\frac{d\sigma}{dt}$$

$$EK\sigma^n \, dt = -d\sigma$$

$$dt = -\frac{1}{EK}\frac{d\sigma}{\sigma^n}$$

Integrating

$$t = -\frac{1}{EK}\left(\frac{\sigma^{-n+1}}{-n+1}\right)$$

$$= \frac{1}{EK(n-1)}(\sigma^{-n+1})$$

Stress falls from σ_0 to σ_t in time t so we have

$$t = \frac{1}{EK(n-1)}\left(\frac{1}{\sigma_t^{n-1}} - \frac{1}{\sigma_0^{n-1}}\right)$$

Example 7.14

A thin-walled pressure vessel made from an austenitic high-temperature alloy has an inner diameter of 45 cm. The vessel operates at 815 °C. Find the allowable internal pressure if the maximum allowable increase in diameter is 5 mm over a 2 year period. Assume steady-state conditions and use the following data for creep tests at 815 °C on the alloy.

Stress ($MN \, m^{-2}$)	69	138
Minimum creep rate ($\%h^{-1}$)	8×10^{-5}	25×10^{-3}

Minimum creep rate, $\dot{\varepsilon}$, and stress, σ, are related by the expression

$$\dot{\varepsilon} = B\sigma^n$$

where B and n are constants at constant creep test temperature. Applying this relationship to the creep data for this alloy, we have

$$8 \times 10^{-5} = B(69)^n$$
$$25 \times 10^{-3} = B(138)^n$$

whence $n = 8.3$ and $B = 8.7 \times 10^{-20.3} \%h^{-1}$.

$$\text{Creep rate of pressure vessel} = \frac{(0.5/45) \times 100}{24 \times 365 \times 2}$$

$$= 6.34 \times 10^{-5} \%h^{-1}$$

Using the $\dot{\varepsilon}$-σ relationship, we have

$$6.34 \times 10^{-5} = 8.7 \times 10^{-20.3} \times \sigma_p^{8.3}$$

where σ_p = allowable internal pressure to accommodate this creep strain rate. Whence $\sigma_p = 66.7$ MN m^{-2}.

Example 7.15

A material which complies with the Arrhenius equation

$$\text{rate} = A \exp(-B/T)$$

where A and B are material constants, T = absolute temperature, is tested at a stress of 100 MN m^{-2}. At 900 K the minimum creep strain rate is 10^{-4} s^{-1} and at 750 K the minimum creep strain rate is 7.5×10^{-9} s^{-1}. Calculate the maximum working temperature of the material if the minimum creep strain rate is 10^{-7} s^{-1}.

$$\text{rate} = A \exp(-B/T) \qquad (1)$$
$$\log \text{rate} = \log A - \frac{B}{2.3T}$$

At 900 K, rate = 10^{-4} s^{-1}; at 750 K, rate = 7.5×10^{-9} s^{-1}. Putting these numerical values in equation 1

$$-4 = \log A - \frac{B}{2.3 \times 900}$$

$$-8.125 = \log A - \frac{B}{2.3 \times 750}$$

whence $\log A = 16.63$ and $B = 4.271 \times 10^4$. Thus at rate = 10^{-7} s^{-1}, we have

$$-7 = 16.63 - \frac{4.271 \times 10^4}{2.3T}$$

whence T = 786 K.

Example 7.16

A magnesium-aluminium alloy is used to make containers for fuel
elements in nuclear reactors. The useful life of the container can
be considered to have ended when a (tensile) creep strain of 1% is
achieved. The life of the fuel elements is 3 years. Experiments
show that at the stresses involved, the containers will last 5 years
at 450 °C. Find whether the life of the containers will be less
than the life of the fuel elements at 500 °C. The activation energy
for creep of the alloy is 133.76 kJ mole^{-1}. The molar gas constant
may be taken as 8.36 J mole^{-1} K^{-1}.
(London)

Putting the respective numerical values in the Larsen-Miller
parameter at 450 °C

$$\left(\frac{E}{R}\right)_\sigma = T (\ln A + \ln t - \ln \varepsilon)$$

we have

$$\frac{133.76 \times 10^3}{8.36} = 723 [\ln A + \ln(5 \times 365 \times 24 \times 3600) - \ln(0.01)]$$

whence ln A = - 1.35.

At 500 °C, the parametric equation is

$$\frac{133.76 \times 10^3}{8.36} = 773 [- 1.35 + \ln t - \ln 0.01]$$

whence t = 1.2 years. Fuel elements last for 3 years. Thus at
500 °C the life of the containers will be less than the life of the
fuel elements.

Example 7.19

Stress relaxation is much more significant in polymers than in metals
and may be expressed by the empirical equation

stress, $\sigma = A + B \log t$

where A and B are constants and t = time (s). If a piece of poly-
sulphide rubber dental impression material requires a stress of
14 MN m^{-2} to produce a strain of 0.02 in 15 s, what stress will be
required to maintain the same strain after 5 h? The constant B in
the quoted equation is -2 MN m^{-2}.

Putting the values for σ, B and t in the equation $\sigma = A + B \log t$, we have

$$14 = A - 2 \log 15$$

whence $A = 16.35$.

After 5 h,

$$\text{stress} = 16.35 - 2 \log(5 \times 3600) = 7.84 \text{ MN m}^{-2}$$

Example 7.18

Given the following creep test results on an alloy, calculate the material constant (expressed as a log term), assuming that secondary creep can be described by a rate-process theory.

Temperature (°C)	650	675
Secondary creep strain	4×10^{-6}	1.03×10^{-2}
Time (s)	10^3	10^6
Molar gas constant, R	$= 8.314$ J mol^{-1} K^{-1}	

Creep strain rate $\dot{\varepsilon}$ = A exp($-Q/RT$)

$$\log \dot{\varepsilon} = \log A - \frac{Q}{2.3RT}$$

so $\log(4 \times 10^{-9}) = \log A - \dfrac{Q}{2.3 \times 8.314 \times 923}$ \hfill (1)

$$\log(1.03 \times 10^{-8}) = \log A - \frac{Q}{2.3 \times 8.314 \times 948} \qquad (2)$$

whence $Q = 28 \times 10^4$ J mol^{-1}. Substituting for E in equation 2, we have $\log A = 7.45$.

PROBLEMS

(1) A low-pressure joint in a pipeline is made by a lead gasket 3 mm thick between steel flanges. The total area of the interfaces between the lead and the steel is 2300 mm^2 and the flanges are bolted by four 5 mm diameter steel bolts.

Taking the free length of the bolts as 40 mm, find the time taken for the interfacial pressure to have reduced to one-half the initial value, which is calculated to be 4.5 MN m^{-2}. The joint is at room temperature. Young's modulus for steel = 206 GN m^{-2} and the secondary creep rate can be taken as

$$\dot{\varepsilon} = 4.2 \times 10^{-14} \sigma^{7.68} \text{ min}^{-1}$$

where σ = applied stress in MN m^{-2}. Distortion of the flanges or adjacent parts of the pipe may be ignored in comparison with that of the gasket and the elastic stretch of the bolts. Neglect the elastic and primary creep deformation of the lead.

[1268 h]

(2) The secondary creep characteristics of a certain material at a constant temperature are found by measurement to be as follows.

Stress $(MN\ m^{-2})$	70	63	56	49	42	35
Creep strain, $\dot{\varepsilon}\ \dfrac{\times 10^5}{(day^{-1})}\ m/m$	5.20	2.48	1.09	0.43	0.15	0.04

A 500 m length of rod of the same material is suspended vertically from one end. Find the over-all extension during a 5 year period due to secondary creep resulting from self-weight only. Density of material = 8600 kg m^{-3}.
(London)

[1.36 m]

(3) It has been suggested that a turbine disc in an aircraft gas turbine engine might be fabricated from a particular high-temperature creep-resistant nickel-based alloy. The maximum stress to which the disc will be subjected is 300 MN m^{-2}, and this will occur near the rim of the disc during take-off when the rim temperature is 765 °C. From the given creep data (figure 7.12) determine whether this suggestion is a reasonable one if the creep strain in the disc is not to exceed 0.1% in 1000 h. It can be assumed that primary creep is insignificant and that the only significant creep strain occurs during the take-off events.
(London)

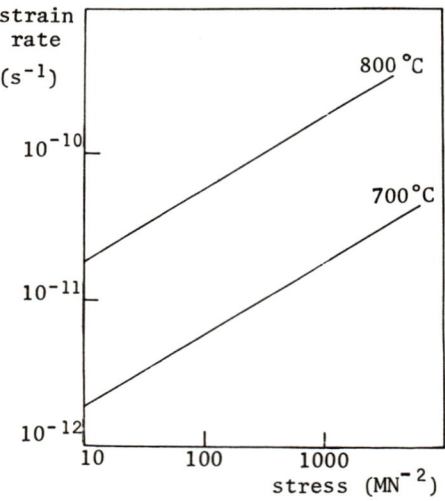

Fig. 7.12

[Suggestion is reasonable because creep strain calculated using data from figure 7.12 is 1.8 × 10^{-4} in 1000 h, which is less than the limit of 10^{-1}% in 1000 h]

(4) Determine the increase in diameter, due to creep, of a thin steel ring 0.75 m in diameter, running at 3000 rev/min for 10 000 hours at

98

a temperature of 500 °C. The circumferential stress at any radius r is given by $\sigma = \rho r^2 \omega^2$ where ρ is the material density and ω is the angular speed of rotation. Take $\rho = 7800$ kg m^{-3} and the creep rate of steel at 500 °C as $\dot{\varepsilon} = 6.25 \times 10^{-13}\sigma^{2 \cdot 7}$ h^{-1}, with σ being the stress in MN m^{-2}.

[1.46 mm]

(5) The creep tests of phosphor bronze ar 250 °C gave a creep rate in mm per mm of 4×10^{-5} per hour occurring at a stress of 148 MN m^{-2} and 4×10^{-7} per hour at 69 MN m^{-2}. Determine the stress to produce a creep strain of 1% in 100 000 h.

[55 MN m^{-2}]

(6) A proposed turbine blade material shows a steady-state creep rate of 10^{-4} mm mm^{-1} s^{-1} at 500 °C under a stress of 138 MN m^{-2}. At what temperature will this rate be doubled at the same stress in this material if the activation energy for the creep process is 168 kJ mole^{-1}? Also calculate the viscosity of this material at this temperature and stress.

[521 °C; 23×10^{10} N s m^{-2}]

(7) A steel bolt clamping two rigid plates is held together at a temperature of 538 °C. Tests at this temperature gave n = 3.0 and a creep rate of 2.8×10^{-8} h^{-1} at a stress of 28 MN m^{-2}. In the relationship creep rate, $\dot{\varepsilon} = B\sigma^n$, where stress, σ, is in MN m^{-2}. If the bolt is initially tightened to a stress of 69 MN m^{-2}, determine the stress after 5 years have elapsed.

[6.56 MN m^{-2}]

(8) The following data were obtained during a creep test of an asphalt paving mixture under a tensile stress of 0.69 MN m^{-2} and a temperature of 15 °C.

Time (s)	100	200	300
Creep strain (%)	0.60	0.80	1.00

The data refer to the minimum creep period. Calculate the viscosity of the mixture.

[1.15×10^{10} N s m^{-2}]

8   FATIGUE FAILURE

8.1 INTRODUCTION

Many components in service are subjected to high-frequency cyclic
or fluctuating stresses. When this happens, the component fails in
a brittle fashion at stresses which are very much less than the
maximum strength of the material under static loading conditions.
This type of failure, known as fatigue failure, occurs in all types
of materials except glasses and is easily the commonest type of
failure of engineering structures in service.

   The simplest type of fatigue test is one in which a round
specimen is subjected to completely reversed bending loads. A
series of identical specimens are tested to fracture, starting at a
high cyclic stress level and progressively reducing this on sub-
sequent specimens. The results (figure 8.1) give the fatigue limit
(in ferrous materials only) as the stress level at which the S-log N
curve becomes horizontal. For non-ferrous materials, the endurance
limit (analogous to the fatigue limit) is the stress value
corresponding to an arbitrarily large number of stress cycles,
usually 5×10^7.

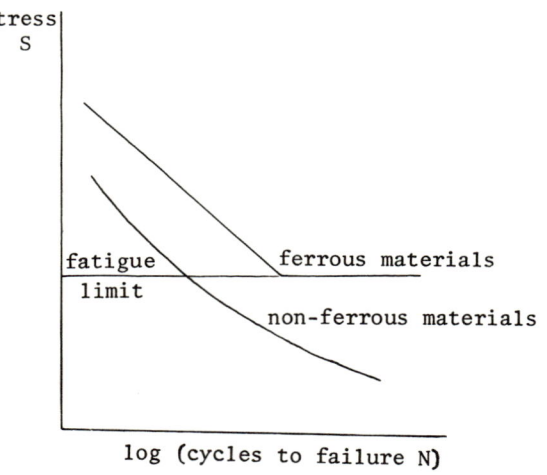

Figure 8.1

A typical cyclic stress cycle is shown in figure 8.2.

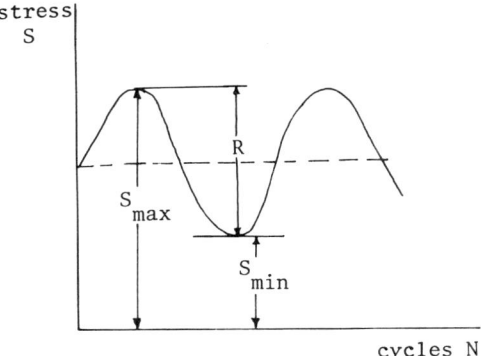

Figure 8.2

8.2 EMPIRICAL FORMULAE

The most satisfactory empirical formula embodying experimental
fatigue test data for steels is due to Gerber and may be written as

$$S_{max} = \frac{R}{2} + \sqrt{\left(S_u^{\,2} - nRS_u\right)}$$

where S_{max} = maximum stress during each cycle at the fatigue limit,
R = stress range, S_u = nominal static ultimate tensile stress and n
is a material constant (= 1.5 for mild steel and 2.0 for high tensile
steel). Gerber's formula may be written

$$S_m = \sqrt{(S_u^{\,2} - nRS_u)}$$

where S_m is the mean stress, or

$$R = \frac{S_u}{n} \left(1 - \frac{S_m^{\,2}}{S_u^{\,2}}\right)$$

where R is the stress range. Goodman suggested a simple straight
line law relating stress range and mean stress

$$R = \frac{S_u}{n} \left(1 - \frac{S_m}{S_u}\right)$$

Now S_m and S_r, the variable stress, are defined as follows

$$S_m = \tfrac{1}{2}(S_{max} + S_{min})$$

and $\quad S_r = \tfrac{1}{2}(S_{max} - S_{min}) = \dfrac{R}{2}$

101

The most commonly used empirical fatigue failure equations, defining the relationship between the mean and variable stresses, are written as

(a) Gerber parabolic equation

$$\frac{S_r}{S_e} + \left(\frac{S_m}{S_u}\right)^2 = 1 \tag{8.1}$$

(b) Modified Goodman equation

$$\frac{S_r}{S_e} + \frac{S_m}{S_u} = 1 \tag{8.2}$$

(c) Soderberg relationship

$$\frac{S_r}{S_e} + \frac{S_m}{S_{yp}} = 1 \tag{8.3}$$

where S_e = fatigue strength for complete stress reversal, S_{yp} = static yield stress.

8.3 THEORIES OF FAILURE FOR COMBINED STRESSES

In these theories, the assumptions and limitations are as follows.

(1) The directions of the principal stresses do not change during loading.
(2) The maximum and minimum values of each principal stress occur at the same time.
(3) The values of the maximum and minimum stress components do not vary with time.
(4) The uniaxial fatigue strength relation is assumed to be represented by the Soderberg equation.
(5) The material is assumed to be isotropic and homogeneous.
(6) Each theory is based on a specific physical concept which defines the limiting value of some stress function.

Maximum stress theory

The principal stresses S_1, S_2 and S_3 vary such that their maximum values are $S_1{'}$, $S_2{'}$ and $S_3{'}$ and their mean values $S_1{''}$, $S_2{''}$ and $S_3{''}$. Then, from equation 8.3, with $p = S_e/S_{yp}$, we have

$$S_1{'} = (1 - p)S_1{''} + S_e$$

$$S_2{'} = (1 - p)S_2{''} + S_e \tag{8.4}$$

$$S_3{''} = (1 - p)S_3{''} + S_e$$

Depending on the relative values of the mean principal stresses, one of equations 8.4 defines fatigue failure by the maximum stress theory.

Maximum Shear Stress Theory

This theory is based on the uniaxial fatigue relation

$$S_{max} = (1 - p)S_m + S_e$$

and assumes that the maximum shear stress is the limiting stress function defining failure. The theory can be stated as

$$S_1' = S_2' + (1 - p)(S_1'' - S_2'') + S_e$$

$$S_2' = S_3' + (1 - p)(S_2'' - S_3'') + S_e \qquad (8.5)$$

$$S_3' = S_1' + (1 - p)(S_3'' - S_1'') + S_e$$

The governing equation here will depend on the relative values of the six principal stresses, and all of the equations 8.5 must be considered in defining failure.

Octahedral Shear Stress Theory

This theory is based on the assumption that failure by fatigue occurs when the maximum octahedral shear stress under cyclic combined stresses equals the maximum octahedral shear stress for cyclic uniaxial stress at failure, provided that the mean value of the octahedral shear stress for the case of combined stresses equals the mean value of this stress under uniaxial cyclic loads.

For an element subjected to cyclic triaxial stresses, the maximum and mean values of the octahedral shear stresses are

$$S_{so}' = \frac{\sqrt{2}}{3}(S_1'^2 + S_2'^2 + S_3'^2 - S_1'S_2' - S_2'S_3' - S_3'S_1')^{\frac{1}{2}}$$

and $$S_{s0}'' = \frac{\sqrt{2}}{3}(S_1''^2 + S_2''^2 + S_3''^2 - S_1''S_2'' - S_2''S_3'' - S_3''S_1'')^{\frac{1}{2}}$$

For fluctuating simple tensile stresses the values of the octahedral shear stresses corresponding to the maximum and mean values at failure can be determined by putting $S_1' = S_{max}$; $S_1'' = S_m$; $S_2' = S_2'' = S_3' = S_3'' = 0$ in equation 8.6. Then

$$S_{s0}' = \frac{\sqrt{2}}{3} S_{max}$$

and $$S_{s0}'' = \frac{\sqrt{2}}{3}S_m$$

From the Soderberg relationship (equation 8.3)

$$S_{s0}' = (1 - p)S_{s0}'' + \frac{\sqrt{2}}{3} S_e \qquad (8.7)$$

From equations 8.6 and 8.7 we have

$$\sqrt{[(S_1{}')^2 + (S_2{}')^2 + (S_3{}')^2 - (S_1{}'S_2{}' + S_2{}'S_3{}' + S_3{}'S_1{}')]}$$
$$-(1 - p)\sqrt{[(S_1{}'')^2 + (S_2{}'')^2 + (S_3{}'')^2 - (S_1{}''S_2{}'' + S_2{}''S_3{}''}$$
$$+ S_3{}''S_1{}'')]} = S_e \tag{8.8}$$

8.4 INFLUENCE OF STRESS CONCENTRATION ON FATIGUE STRENGTH

A measure of the influence of notches on fatigue test results can be
defined by the fatigue notch reduction factor, k_f, which is the ratio
of the fatigue strength of unnotched specimens at N cycles to the
fatigue strength of notched specimens at the same number of cycles.
The relationship between the fatigue notch factor and the static
stress concentration factor, k_t, is through the notch sensitivity
factor, q, and is given as

$$q = \frac{k_f - 1}{k_t - 1}$$

or $\quad k_f = q(k_t - 1) + 1$

and the estimated fatigue notch factor k_{tf} is given as

$$k_{tf} = q(k_t - 1) + 1$$

Based on the Soderberg equation, the failure relationship for
notched specimens with uniaxially steady stresses superimposed on
reversed stresses is

$$\frac{k_{tf}S_r}{S_e} + \frac{S_m}{S_{yp}} = 1 \tag{8.9}$$

indicating that a stress concentration factor is only applied to the
variable part of the stress. This conclusion is based on the
assumption that on stressing beyond the yield point of the material
plastic deformation occurs. A more logical and accurate analysis
takes this into account and so equation 8.9 becomes

$$\frac{k_{tf}S_r}{S_e} + \frac{k_p S_m}{S_{yp}} = 1 \tag{8.10}$$

where k_p = static plastic stress concentration factor.

8.5 UTILISATION OF FATIGUE PROPERTIES IN DESIGN

In considering the design of members for fatigue loading it is first
necessary to establish working or design stress relationships for
both simple and combined stresses based on failure relationships for

such stresses. No standard basis for obtaining working fatigue stress relationships exists. In this section, we shall limit the consideration to the conservative Soderberg relationship which assumes equal strengths in uniaxial tension and compression. Also, it is assumed that (a) there are no directional effects, (b) the material is isotropic, (c) stress amplitudes are constant, and (d) the state of stress is uniaxial or biaxial. With these conditions, failure for uniaxial stresses is defined by equation 8.10. A working stress relationship is obtained from equation 8.10 by replacing S_e and S_{yp} by allowable values (S_e/N_e) and (S_{yp}/N_y), where N_e and N_y are selected factors of safety. Then the working stress relationship for uniaxial stresses becomes

$$k_{tf} S_r + k_p \left(\frac{S_e}{S_{yp}}\right) \left(\frac{N_y}{N_e}\right) S_m = \frac{S_e}{N_e} \qquad (8.11)$$

Example 8.1

A rotating stepped shaft is subjected to a constant bending moment. The shaft is made of a carbon steel with an ultimate tensile strength of 450 MN m^{-2} and a limit of endurance in flexure of 220 MN m^{-2} (symmetrical cycle). The outer and inner diameters of the shaft are 100 mm and 80 mm respectively, the fillet has a radius of 10 mm (figure 8.3). Determine the maximum allowable value of the bending moment. Take the factor of safety, with respect to the endurance limit, to be 2.

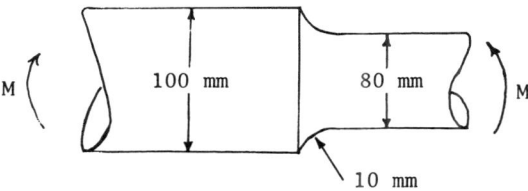

Figure 8.3

Let M = constant bending moment, S_u = ultimate tensile strength, S_e = limit of endurance in flexure, k = factor of safety referred to the endurance limit, r = fillet radius, D = outer diameter, d = inner diameter. The endurance limit of a component may be obtained from

$$S_e = \frac{S_e{'}}{k_{tf} k_s}$$

where $S_e{'}$ = endurance limit obtained experimentally on a scale model of the component under identical conditions, k_{tf} = stress concentration factor for the model and k_s = scale factor.   Now

$$k_{tf} = q(k_t - 1) + 1$$

The value of k_t may be obtained from standard tables (appendix III). At $r/d = 10/80 = 0.125$, $k_t = 1.5$. From figure 8.4 at this value of k_t, q for a steel with an ultimate tensile strength of 450 MN m^{-2} is 0.36. Thus $k_{tf} = 1.18$.

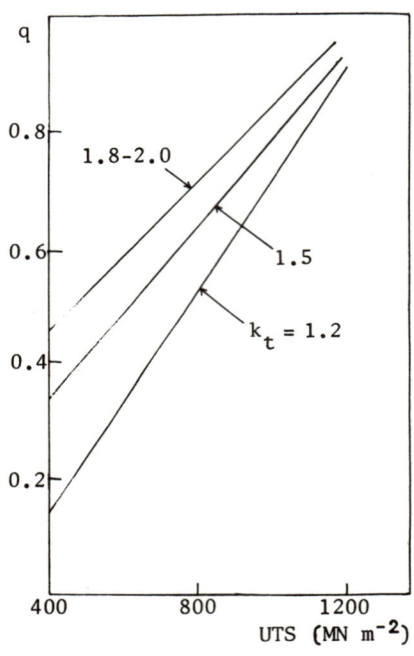

Fig. 8.4

The value of k_s is determined from figure 8.5: at d = 80 mm, k_s = 1.56, The endurance limit of the shaft in a symmetrically vary-ing stress cycle is

$$S_e = \frac{220}{1.18 \times 1.56} = 119 \text{ MN m}^{-2}$$

$$\text{allowable stress} = \frac{S_e}{k} = \frac{119}{2} = 59.5 \text{ MN m}^{-2}$$

The allowable value of the bending moment, M, is determined from the condition of limiting stress as

$$\sigma_{max} = \frac{M_{max}\, y}{I} = \frac{32 M_{max}}{\pi d^3} < \frac{S_e}{k}$$

$$M_{max} = \frac{59.5 \times \pi \times (0.080)^3}{32} = 3 \text{ kN m}$$

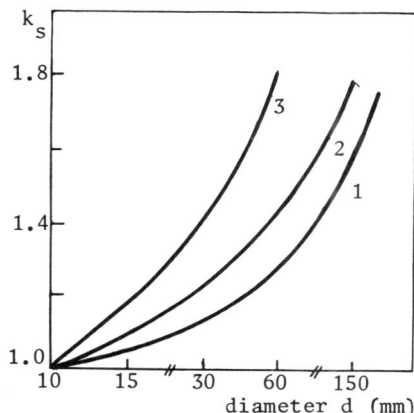

Fig. 8.5

Example 8.2

A tensile member 108 mm by 25 mm in cross-section with a central hole of 6 mm diameter, is made of annealed steel. The member is subjected to a fluctuating axial load which varies from a minimum value of P/2 to a maximum value of P. What value of P will produce failure in 10^6 cycles?

The yield strength of the material is 420 MN m^{-2} and the fatigue strength for complete reversal is 315 MN m^{-2} for 10^6 cycles. Determine the notch fatigue factor using the notch sensitivity factor data given in figure 8.6, and a static stress concentration factor of 3.0. Assume that the static plastic stress-concentration factor in tension is 1.5 and that the failure is defined by the Soderberg relationship.

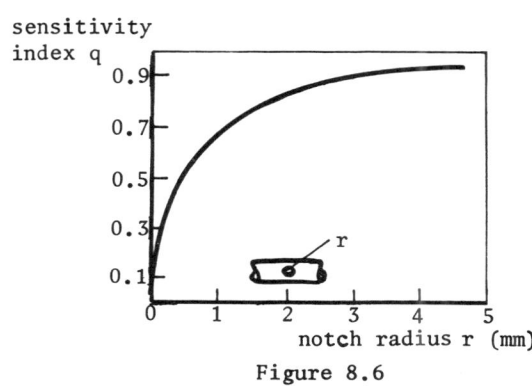

Figure 8.6

By the Soderberg equation, failure is defined by

$$k_{tf}S_r + pk_pS_m = S_e \tag{1}$$

Fatigue notch factor k_{tf} is defined as

$$k_{tf} = q(k_t - 1) + 1$$

From figure 8.6 for a hole of radius 3 mm, q = 0.9 so

$$k_{tf} = 0.9(3 - 1) + 1 = 2.8$$

Net area, $A = (108 - 6) \times 25 = 2550$ mm^2 = 2.55×10^{-3} m^2

Now $S_r = \dfrac{P - P/2}{2A} = \dfrac{P}{4A}$

$S_m = \dfrac{P + P/2}{2A} = \dfrac{3P}{4A}$

and $P = \dfrac{315}{420} = 0.75$

Then, from equation 1

$$\frac{2.8P}{4 \times 2.55 \times 10^{-3}} + \frac{0.75 \times 1.5 \times 3P}{4 \times 2.55 \times 10^{-3}} = 315 \times 10^6$$

whence P = 520 kN.

Example 8.3

A circular steel shaft having a transverse oil hole is subjected to a torsional load which varies from -100 N m to +400 N m (i.e. values in opposite senses). Determine the necessary shaft diameter assuming that the hole causes a fatigue strength reduction factor of 1.75. Assume the following properties for the steel: ultimate shear strength = 200 MN m^{-2}, fatigue limit in reversed torsion = 130 MN m^{-2}.

Using the Soderberg relationship

$$\frac{k_{tf}S_r}{S_e} + \frac{S_m}{S_{yp}} = 1 \tag{1}$$

Now $S_r = \dfrac{500 \times 32}{2 \times 2 \times \pi d^3} = \dfrac{1273}{d^3}$

and $S_m = \dfrac{32 \times 300}{2 \times 2\pi d^3} = \dfrac{764}{d^3}$

So equation 1 becomes

$$\frac{1.75 \times 1273}{d^3 \times 130 \times 10^6} + \frac{764}{d^3 \times 200 \times 10^6} = 1$$

whence d = 28 mm.

Example 8.4

A tension member is made of a material which has an ultimate tensile strength of 470 N mm^{-2} and a fatigue limit (corresponding to 10^6 cycles for completely reversed stress cycle) of 185 N mm^{-2}. The cross-sectional area of the member is 320 mm^2. Stress concentration factor, k_t, equals notch sensitivity factor and is of value 1.8. The member is subjected to a range of cyclic loading from a minimum load of P_{min} of 18 000 N to a maximum load P_{max}. Determine the maximum load required to cause failure in 10^6 cycles of loading, using Goodman's relationship.

The modified Goodman relationship (taking into account stress concentration) is

$$\frac{k_{tf} S_r}{S_e} + \frac{k_p S_m}{S_u} = 1$$

or $\quad k_{tf} S_r + p k_p S_m = S_e$ $\hfill$ (1)

Now $k_{tf} = q(k_t - 1) + 1$ which in this case is equal to $1.8(1.8 - 1) + 1 = 2.44$ and k_p is assumed equal to unity.

$$S_r = \frac{P_{max} - P_{min}}{2A}$$

and $\quad S_m = \dfrac{P_{max} + P_{min}}{2A}$

whence A = 0.32×10^{-3} m^2 and p = S_e/S_u = 185/470 = 0.39. Equation 1 thus becomes

$$\frac{2.44}{2A}(P_{max} - P_{min}) + \frac{0.39}{2A}(P_{max} + P_{min}) = 185 \times 10^6$$

P_{min} = 18 kN, whence P_{max} = 55 kN.

Example 8.5

A member of circular cross-section with a diameter of 50 mm is subjected to an axially applied static tensile load, P_m, of 90 kN. What completely reversed load can be superimposed on P_m at a distance

of 12.5 mm from the centroid of the cross-section so that failure
will not occur for 10^6 cycles? The fatigue strength for completely
reversed stress at 10^6 cycles is 207 MN m^{-2}, and the static yield
strength is 275 MN m^{-2}. Use the Soderberg relationship.

The critical variable and mean stresses are

$$S_r = \frac{M_r r}{I} = \frac{4M_r}{\pi r^3}$$

and $\quad S_m = \frac{P_m}{A} = \frac{P_m}{\pi r^2}$

where r = radius of the critical cross-section. Putting these values
of S_r and S_m in the Soderberg relationship

$$\frac{S_r}{S_e} + \frac{S_m}{S_{yp}} = 1$$

we have (since $p = S_e/S_{yp}$)

$$\frac{4M_r}{\pi r^3} + \frac{pP_m}{\pi r^2} = S_e$$

Now $M_r = P_r e$ (e = eccentricity) so we have

$$P_r = \frac{\pi r^3 S_e}{4e} - \frac{pP_m r}{4e}$$

$$= \frac{1}{4 \times 12.5}(\pi \times 25^3 \times 207 - 0.75 \times 90 \times 10^3 \times 25)$$

$$= 169 \text{ kN}$$

Example 8.6

A fluctuating axial tensile load varying from 180 kN to 440 kN is
applied to a bolt. The material used has a tensile yield strength of
420 MN m^{-2} and a fatigue strength of 280 MN m^{-2}. Determine the
required cross-sectional area of the bolt using the Soderberg
relationship. Static stress concentration factor = 1.5, fatigue
stress concentration factor = 1.0, static factor of safety = 2.0,
fatigue factor of safety = 4.0.

The variable and mean stresses are

$$S_r = \frac{P_r}{A} = \frac{130 \times 10^3}{A} \text{ Nm}^{-2}$$

and $S_m = \dfrac{P_m}{A} = \dfrac{310 \times 10^3}{A}$ Nm^{-2}

The Soderberg relationship (with stress concentration effects) is

$$\frac{k_{tf}S_r}{S_e} + \frac{k_p S_m}{S_{yp}} = 1 \tag{1}$$

With factors of safety N_e and N_y, $S_e{}^{\prime} = Se/Ne$ and $S_{yp}{}^{\prime} = S_{yp}/N_y$. Hence equation 1 becomes

$$k_{tf}S_r + \frac{k_p S_m S_e N_y}{S_{yp} N_e} = \frac{S_e}{N_e}$$

Putting in the various values for the parameters, we have

$$\frac{1.5 \times 130 \times 10^3}{A} + \frac{310 \times 10^3 \times 280 \times 2}{A \times 420 \times 4} = \frac{280 \times 10^6}{4}$$

whence A = 4.25 $\times$ 10^{-3} m^2 = 4250 mm^2.

Example 8.7

A pressure cylinder with 100 mm inside diameter and 6 mm wall thickness is subjected to internal pressure that varies from $-p/4$ to p. The steel has a fatigue limit of 41.4 MN m^{-2} in completely reversed bending. Under a repeated stress cycle of zero to a maximum stress in tension the same steel can withstand 62 MN m^{-2} without failure in 10^7 cycles. Calculate the maximum internal pressure that can be withstood by the cylinder without fatigue failure before 10^7 cycles, using the Soderberg relationship.

The variable and mean stresses are

$$S_r = \left(\frac{P_{max} - P_{min}}{2}\right)\frac{d}{2t} = \frac{5 \times 100}{4 \times 2 \times 12} = 5.208p \text{ N mm}^{-2}$$

$$S_m = \left(\frac{P_{max} + P_{min}}{2}\right)\frac{d}{2t} = \frac{3 \times 100}{4 \times 2 \times 12} = 3.125p \text{ N mm}^{-2}$$

Putting these values in the Soderberg relationship

$$\frac{S_r}{S_e} + \frac{S_m}{S_{yp}} = 1$$

we have

$$\frac{5.208p}{41.4} + \frac{3.125p}{62} = 1$$

whence p = 5.67 MN m^{-2}.

Example 8.8

A bolt 3000 mm^2 in cross-section is made of a material with a fatigue strength at 10^7 cycles of 200 MN m^{-2} and an ultimate tensile strength of 300 MN m^{-2}. The bolt is subjected to a static mean tensile load of 180 kN. Using Goodman's relationship calculate the reversed fatigue load which will produce failure in 10^7 cycles.

If P_m and P_r are the mean and variable loads, then the mean and variable stresses are $S_m = P_m/A$ and $S_r = P_r/A$. By Goodman's relationship

$$\frac{S_r}{S_e} + \frac{S_m}{S_u} = 1 \tag{1}$$

Placing values of S_r and S_m from above in equation 1 we have

$$\frac{P_r}{S_e} + \frac{P_m}{S_u} = A$$

or $\quad P_r = S_e \left(A - \frac{P_m}{S_u} \right)$

Inserting the numerical values for the various terms, we have

$$P_r = 200 \left(3000 \times 10^{-6} - \frac{180 \times 10^3}{300 \times 10^6} \right) \text{ MN}$$

whence $P_r = 480$ kN.

Example 8.9

Acylindrical vessel 0.762 m internal diameter and having a wall thickness of 6 mm is subjected to internal pressure fluctuating from P/2 to P. The fatigue limit of the material at 10^7 cycles under reversed direct stress is 230 MN m^{-2} and the static yield stress is 310 MN m^{-2}. Assuming the octahedral shear theory of failure and using the Soderberg relationship determine a nominal value for P to give an endurance of 10^7 cycles. The radial stress through the wall thickness may be taken as zero.

The maximum and mean values of the principal stresses are

$$s_1' = \frac{P'd}{2t} = \frac{P \times 762}{2 \times 6} = 63.5P$$

$$s_1'' = \frac{P''d}{2t} = \frac{P \times 762}{2 \times 2 \times 2 \times 6} = 15.88P$$

112

$$s_2' = \frac{P'd}{4t} = \frac{P \times 762}{4 \times 6} = 31.75P$$

$$s_2'' = \frac{P''d}{4t} = \frac{P \times 762}{16 \times 6} = 7.94P$$

Putting these values in the expression for the octahedral shear theory of failure

$$\sqrt{[(S_1')^2 + (S_2')^2 - S_1'S_2']} - 1 - \left(\frac{S_e}{S_{yp}}\right)\sqrt{[(S_1'')^2 + (S_2'')^2 - S_1''S_2'']} = S_e$$

we have

$$54.99P - (1 - 0.74)13.75P = 230$$

whence $P = 4.47$ MN m^{-2}.

Example 8.10

On a diagram of semi-range of stress against mean stress sketch the Goodman, Gerber abd Soderberg relationships.

A horizontal steel cantilever 0.7 m long and of rectangular section 0.1 m wide and 0.2 m deep is subjected to a fluctuating end load acting vertically downwards. The load varies sinusoidally from a minimum value of 3×10^4 N to a maximum value P_{max}. The steel has a fatigue strength of 200 MN m^{-2} at 10^6 cycles in an alternating test and a tensile yield strength of 300 MN m^{-2}. Use the Soderberg relationship to find the value of P_{max} for a design life of 10^6 cycles.
(Cambridge)

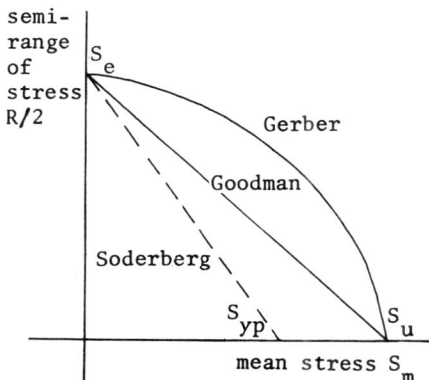

Figure 8.7

The required graph is shown in figure 8.7.

The variable and mean stresses for the critical element of the beam beam are

$$S_r = \frac{M_r C}{I} = \frac{(M_{max} - M_{min})}{2I}C$$

$$S_m = \frac{M_m C}{I} = \frac{(M_{max} + M_{min})}{2I}C$$

where C = distance of the neutral axis from the top of the section. Placing the values of S_r and S_m from above in the Soderberg relationship

$$\frac{S_r}{S_e} + \frac{S_m}{S_{yp}} = 1$$

we have

$$M_{max} = \frac{2S_e I}{(1 + p)C} + \frac{(1 - p)}{(1 + p)}M_{min} \tag{1}$$

where $p = S_e/S_{yp} = 200/300 = 0.67$. Substituting $M_{max} = P_{max}L$ and $M_{min} = P_{min}L$ in equation 1, and solving for P_{max} gives

$$P_{max} = \frac{2S_e I}{LC(1 + p)} - \frac{(1 - p)}{(1 + p)}P_{min}$$

Putting in the numerical values for the various terms, we have

$$P_{max} = \frac{2 \times 200 \times 10^6 \times 0.1 \times 0.2^3}{12 \times 0.7 \times 0.1 \times 1.67} - \frac{0.33 \times 3 \times 10^4}{1.67} = 222 \text{ kN}$$

Example 8.11

(a) AISI 4340 steel has an ultimate tensile stress of 600 MN m^{-2}. In a fatigue test, it is subjected to a completely reversed stress cycle of ± 180 MN m^{-2}. Calculate the material constant, according to Gerber's law.

(b) If in another fatigue test it is subjected to a repeated stress cycle, what is the maximum stress at the fatigue limit, according to Goodman's law?

(a) Gerber's law is

$$\sigma_{max} = \frac{R}{2} + \sqrt{(\sigma_u^2 - nR\sigma_u)}$$

For completely reversed stress cycle, $R = 2\sigma_{max}$ so the law reduces to

$$n = \sigma_u R$$

whence $n = 600/360 = 1.67$.

(b) For repeated stress cycle, $R = \sigma_{max}$, $M = \sigma_{max}/2$. Goodman's law is

$$R = \frac{\sigma_u}{n} \left(1 - \frac{M}{\sigma_u}\right)$$

or

$$\sigma_{max} = \frac{\sigma_u}{n} \left(1 - \frac{\sigma_{max}}{2\sigma_u}\right)$$

$$\sigma_{max} \left(1 + \frac{1}{2n}\right) = \frac{\sigma_u}{n}$$

In this case

$$\sigma_{max} \left(1 + \frac{1}{2 \times 1.67}\right) = \frac{600}{1.67}$$

whence $\sigma_{max} = 277$ MN m^{-2}.

Example 8.12

It is required that a component, made from an aluminium alloy, should not fail by fatigue in less than 10^7 cycles while it is subjected to a tensile mean stress of 200 N mm^{-2}. Laboratory tests on the material under alternating fatigue conditions show that

(a) the cyclic yield stress of the materials is 400 N mm^{-2}
(b) when the range of cyclic stress, $\Delta\sigma$, is 340 N mm^{-2} the number of cycles to failure, N_f, is 10^8 and when $\Delta\sigma$ is 520 N mm^{-2}, N_f is 10^5

Assuming that the fatigue behaviour of the material, under alternating stress conditions, is given by

$$\Delta\sigma^b N_f = C$$

where b and C are material constants, calculate the permissible amplitude of cyclic tensile stress to which the component may be subjected.
(Cambridge)

For the fatigue behaviour of the material, we have

$$\Delta\sigma^b N_f = C$$

and putting in the values of σ and N_f, we can evaluate b and C

$$340^b \times 10^8 = C$$

$$520^b \times 10^5 = C$$

whence $b = 16.26$ and $C = 1.45 \times 10^{49}$. The permissible amplitude of cyclic tensile stress at 10^7 cycles can now be computed

$$\Delta\sigma^{16.26} \times 10^7 = 1.45 \times 10^{49}$$

whence $\Delta\sigma = 392$ N mm^{-2}

Example 8.13

Crack growth in an alloy takes place according to a relationship of the type

$$\frac{da}{dN} = C(\Delta K)^m$$

where a = half-crack length, N = number of cycles, ΔK = stress intensity factor range, C and m are material constants. Show how the crack-growth relationship can be applied to determine crack-growth under constant nominal stress range if

$$\Delta K = \Delta\sigma\sqrt{(\pi a)}\alpha$$

where $\Delta\sigma$ is the stress range and α is a constant depending on crack geometry.

If $m = 4$, $\Delta K = 5$ MN m$^{-3/2}$ for an initial crack growth rate of 10^{-6} mm per cycle in the material, determine how many cycles will elapse while a flaw of initial length 2 mm grows to 2 cm if $\Delta\sigma = 90$ MN m^{-2}. Assume $\alpha = 1$ for the flaw.
(Cambridge)

Putting in the values of da/dN (in m per cycle) and ΔK (in N m$^{-3/2}$) in the relationship

$$\frac{da}{dN} = C(\Delta K)^m \qquad (1)$$

$$10^{-9} = C(5 \times 10^6)^4$$

whence $C = 16 \times 10^{-37}$. Now

$$\Delta K = \Delta\sigma\sqrt{(\pi a)}\alpha$$

The relevant crack lengths are $a_c = 1$ cm and $a_i = 1$ mm.

From equation 1, we get

$$\int_{0.001}^{0.01} \frac{da}{a^2} = \int_{0}^{N} 16 \times 10^{-37} (90 \times 10^6)^4 \, \pi^2 dN$$

making

$$N = \frac{900}{16 \times 10^{-37} \times 10^{24} \times 90^4 \times \pi^2} \quad \text{cycles}$$

$$= 8.68 \times 10^5 \text{ cycles}$$

Example 8.14

One type of technique which has been applied to the statistical determination of the estimated fatigue limit of materials is the so-called 'staircase' method. Out of 18 specimens tested using this method there were 8 non-failures. The test results are as follows.

Stress Level (MN m^{-2})	Number of Specimens that did not Fail at this Stress Level
317	1
310	2
303	4
296	1

Estimate the mean and standard deviation of the fatigue limit of this material.

In the staircase method, we designate the lowest stress level at which a non-failure was obtained as $i = 0$, the next $i = 0$, etc. The results are set out as follows.

Stress (MN m^{-2})	i	Number of Non-failures, n_i	$i n_i$	$i^2 n_i$
317	3	1	3	9
310	2	2	4	8
303	1	4	4	4
296	0	1	0	0
		$N=8$	$A=11$	$B=21$

Mean value of the fatigue limit is

$$\bar{X} = X_0 + d\left(\frac{A}{N} + \tfrac{1}{2}\right) \text{ MN m}^{-2}$$

Standard deviation of the fatigue limit is

$$S = 1.62d \left(\frac{NB - A^2}{N^2} + 0.029\right) \text{ MN m}^{-2}$$

where X_0 = first stress level, d = stress increment. Then

$$\bar{X} = 296 + 7\left(\frac{11}{8} + \frac{1}{2}\right) \approx 309 \text{ MN m}^{-2}$$

117

$$S = 1.62 \times 7\left(\frac{8 \times 21 - 11^2}{8^2} + 0.029\right) = 8.66 \text{ MN m}^{-2}$$

PROBLEMS

(1) A crankshaft journal contains a transverse drilled oil hole which gives rise to a theoretical stress concentration factor of 3. In service the journal is subjected to alternating bending stresses and the shaft material has an ultimate tensile stress of 900 MN m^{-2}, a fatigue strength of 10^8 cycles in repeated tension of 450 MN m^{-2} and a notch sensitivity index, q, of 0.5. Calculate the maximum value of service stresses remote from the oil hole if the shaft is to withstand at least 10^8 cycles in operation.
(London)
[Assuming stress is from 0 to σ_{max} and the Goodman relationship applies, $\sigma_{max} = 360$ MN m^{-2}]

(2) A propeller shaft of uniform diameter transmits 890 kW at 90 rev/min and the torque is subjected to a variation of ± 40%. The shaft is also subjected to a completely reversed bending moment of 30 kN m. Determine the diameter of the shaft if the over-all factor of safety is 4. Take the yield point of the material to be 345 MN m^{-2} and the endurance limit in cyclic bending = 331 MN m^{-2}.

[18.3 cm]

(3) Estimate the fatigue strength for 10^5 cycles in alternating tension for a 52 mm diameter shaft reduced to 45 mm with a 3.3 mm radius shoulder (figure 8.8). The steel has an unnotched fatigue strength of 262 MN m^{-2} and a tensile strength of 414 MN m^{-2} (see figure 8.9). Take into account a factor of safety of 3.

[55.67 MN m^{-2}]

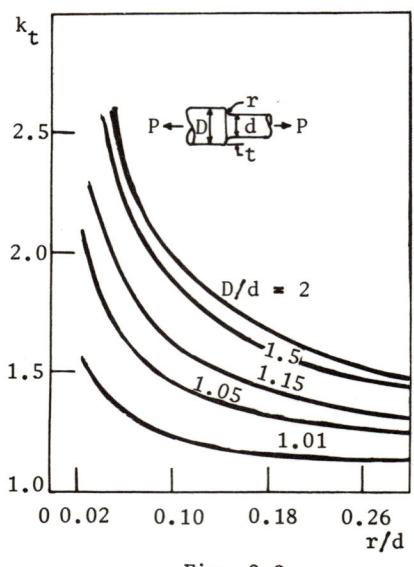

Fig. 8.8

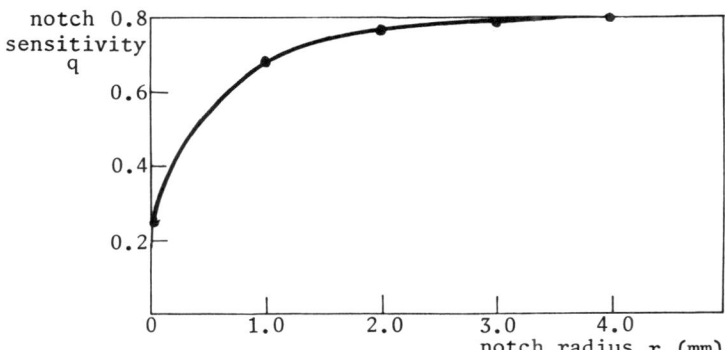

Fig. 8.9

(4) If a closed-coiled helical spring is subjected to a mean shear stress of 310 MN m^{-2}, find the limiting safe value of the maximum stress of a superimposed cyclic stress so that the spring will not fail by fatigue or by permanent set. Assume that the bending stress in the coils can be neglected and that the maximum shear stress theory of failure holds true. Take the torsional yield point to be 1.52 GN m^{-2} and the endurance limit of the spring wire in cyclic stress to be 552 MN m^{-2}.

[220 MN m^{-2}]

(5) The Gerber relationship is sometimes expressed as

$$S_{max} = \frac{S_r}{2} + \sqrt{(S_u^2 - nS_rS_u)}$$

where S_r = stress range, S_u = tensile strength, n is a material constant. Show that this can also be written as

$$S_r = S_r'[1 - (S_m/S_u)^2]$$

where S_r' = stress range for S_m = 0.

(6) Tabulate different (a) chemical or metallurgical and (b) mechanical processes which can be used to increase the fatigue strength of steel by improvement in the surface conditions.

If σ is the depth of the nitrided layer and d is the bar diameter, what is the percentage increase in fatigue strength due to nitriding?

$$\frac{100}{[(d/2\sigma) - 1]}$$

(7) Outline the steps in the calculation of the theoretical reduction in fatigue strength for different kinds of surface finish.

119

If σ = depth of surface scratch, ρ = bottom radius of the surface scratch and C is a constant, derive an expression for the reduction in fatigue strength for a surface scratch.

If the strength reduction factor is 1.2 for a particular surface finish having a value of σ/ρ of 1, what would be the reduction, for the same material, with a coarse finish having a value of σ/ρ of 2?

$$\left[\frac{C\sqrt{(\sigma/\rho)}}{1 + C\sqrt{(\sigma/\rho)}}; \ 1.14\right]$$

9   FRACTURE AND FRACTURE MECHANICS

9.1 BRITTLE FRACTURE

Griffith was the first to offer an explanation for brittle fracture.
He postulated that in a brittle material, cracks propagate when the
released strain energy is just sufficient to provide the surface
energy for the creation of new surfaces. That is elastic strain
energy per unit volume = $\sigma^2/2E$, where σ = stress applied
perpendicular to the crack, E = modulus of elasticity. Very near to
the crack faces the stress falls to zero and very far from the crack
it is unchanged, so we assume that roughly an area of radius C around
the crack is relieved of its elastic energy. Thus for a crack of

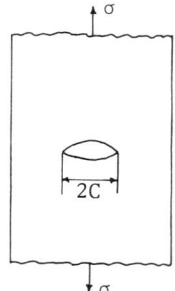

Fig. 9.1

unit width, the total elastic strain energy is $\sigma^2/2E \times \pi C^2 \times 1$.
Properly, however, using Inglis' rigid approach, the strain field
should be integrated from ∞ to the surface of the crack, giving the
total elastic strain energy as

$$U_E = \frac{\sigma^2 \pi C^2}{E}$$

If the surface energy is γ J m^{-2}, then the surface energy for a crack
of length 2C and unit width is

$$U_s = \gamma \times 4C$$

(there are two faces for each crack). Applying Griffith's criterion
for crack propagation ($dU_E/dC = dU_s/dC$), we have

$$\sigma = \sqrt{\left(\frac{2\gamma E}{\pi C}\right)} \text{ N m}^{-2} \tag{9.1}$$

9.2 STRESS CONCENTRATIONS

Stress concentrations generally are associated with geometrical discontinuities, where there are rapid changes in the stress over a small area with the maximum value being considerably higher than the average stress in the full section of the component. Examples of such discontinuities include holes, keyways, splines, gear teeth, etc. The stress distribution so set up is a function of the shape and dimensions of the discontinuity and is expressed in terms of the elastic stress concentration factor, K_t, where

$$K_t = \frac{\text{maximum boundary stress at discontinuity}}{\text{average stress at that cross-section of the component}}$$

Limitations in the theoretical solutions of stress concentrations have meant that experimental results (usually obtained with the aid of photoelasticity) are presented as charts or plots of K_t against geometrical dimensions (see, for example, figure 9.2).

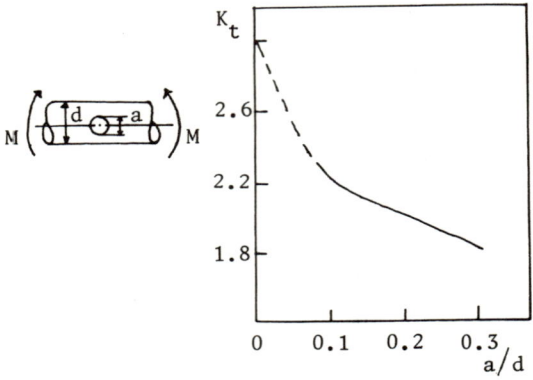

Figure 9.2

9.3 LINEAR ELASTIC FRACTURE MECHANICS

'Real' structures are known to contain either internal or surface flaws. The engineer therefore has to develop criteria to ascertain whether, and at what applied stress, σ, any of these flaws will propagate to failure. Linear elastic fracture mechanics is the analysis of the elastic stresses in the neighbourhood of pre-existent cracks of specified intensity.

The effective stress intensity factor, K, close to the root of a given crack is given as

$$K = \sigma \times \text{a geometrical factor (MN m}^{-3/2})$$

This geometrical factor, known as the compliance function, takes the form $\alpha\sqrt{(\pi C)}$, where C is the effective crack length, and α depends on the dimensions of the specimen and the crack. For example, for a

plate of width W loaded in tension with a centrally located crack of
length 2C

$$\alpha = \sqrt{\left[\frac{W}{\pi C} \tan\left(\frac{\pi C}{W}\right)\right]}$$

A variety of stress conditions may be encountered, the most severe
of these being plane strain (designated mode I and corresponding to
the case where all strain displacements are limited to the two
directions normal to the plane of the crack); the stress intensity
factor is then written K_I and when this reaches some critical value
the elastic strain energy stored in the zone ahead of the crack is
equal to the surface energy required to form additional crack
surfaces if the crack is to extend into the zone. In linear elastic
fracture mechanics the parameter of interest is the fracture toughness, which is the critical value of K_I, and written K_{Ic}.

The concepts of plane strain and plane stress are also important
here and these can easily be defined as shown below.

Plane Strain Conditions	Plane Stress Conditions
$\sigma_x \neq 0,\ e_x \neq 0$	$\sigma_x \neq 0,\ e_x \neq 0$
$\sigma_y \neq 0,\ e_y \neq 0$	$\sigma_y \neq 0,\ e_y \neq 0$
$\sigma_z \neq 0,\ e_z = 0$	$\sigma_z = 0,\ e_z \neq 0$

The relationship of K to σ and C and the geometric factor f(C/W) is
called a K-calibration. These have been determined experimentally
and are available for standard specimens and conditions. For example-
(a) for a 'thumbnail' type surface flaw

$$K_I = 0.78\sigma \sqrt{(\pi C)}$$

and (b) for a bend bar

$$K_I = \frac{PS}{BW^{3/2}}\left[2.9\left(\frac{a}{W}\right)^{1/2} - 4.6\left(\frac{a}{W}\right)^{3/2}\right.$$

$$\left. + 21.8\left(\frac{a}{W}\right)^{5/2} - 37.6\left(\frac{a}{W}\right)^{7/2} + 38.7\left(\frac{a}{W}\right)^{9/2}\right]$$

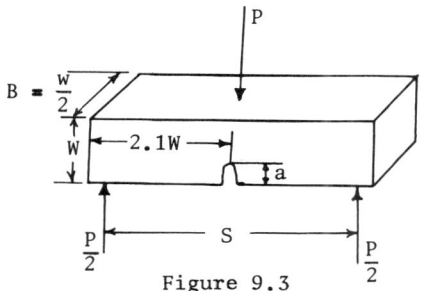

Figure 9.3

Irwin proposed that fracture occurs at a stress corresponding to a critical value of the crack extension force, G_c. This is defined as the force required to produce a unit increase in the crack length. G_c is also sometimes referred to as the strain energy release rate and has units of N m^{-1}.

Irwin proposed that

$$\sigma = \sqrt{\left(\frac{EG_c}{\pi C}\right)}$$

$$G_c = \frac{\pi C \sigma^2}{E}$$

For a sharp elastic crack in an infinitely wide plate, $K_I = \sigma\sqrt{(\pi C)}$, so we have

$$K_I{}^2 = EG_c \text{ (for plane stress conditions)}$$

and $\quad K_I{}^2 = \frac{EG_c}{1 - \gamma^2}$ (for plane strain conditions)

where γ = Poisson's ratio for the material

Example 9.1

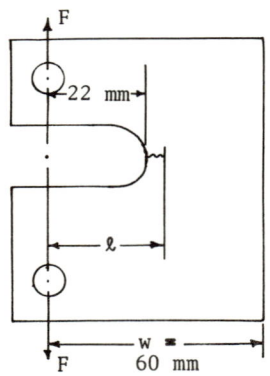

F

22 mm

ℓ

w = 60 mm

F

Fig. 9.4

A thin perspex component is shown in figure 9.4. Due to misuse in service, a fatigue crack 2.7 mm long emanates from the root of the notch. It is required to investigate the possibility of brittle fracture of the component. The only relevant data that are readily available are as follows: Young's modulus of elasticity is 2940 MN m^{-2}; in tests to determine the strain energy release rate, G, on parallel-sided thin strip specimens of perspex containing a single-edge crack of length c subjected to an axial tension P, the following results were obtained

124

c (mm)	4	8	10	12	14
G/P^2 (kN^{-1} mm^{-1})	0.766	2.55	4.85	8.68	15.30

One of these specimens fractured when c was 10.5 mm and P was 347 N.

The stress intensity factor, K, for the geometry of figure 9.4 is given by

$$K = \alpha F \ell^{\frac{1}{2}}/W$$

Where F = applied tensile force and α is a function of ℓ/w. some typical values being

ℓ/w	0.39	0.40	0.42	0.44	0.46
α (mm^{-1})	0.66	0.77	0.90	0.98	1.05

Determine the magnitude of the static force, F, which would cause brittle fracture of the damaged component. (Cambridge)

ℓ = 22 + 2.7 = 24.7 mm, w = 60 mm, ℓ/w = 0.42. So α = 0.90 mm^{-1} (from the given table). A graph of c against G/P^2 is plotted (figure 9.5). At fracture c = 10.5 mm and G/P^2 = 5.60 kN^{-1} mm^{-1}. Now P = 0.347 kN, so

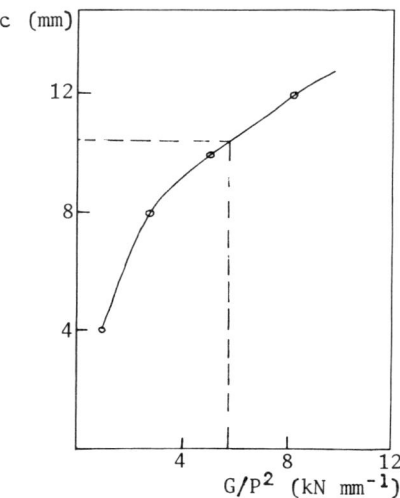

Fig. 9.5

$$G = (0.347)^2 \times 5.60 = 0.674 \text{ kN mm}^{-1}$$

K and G are related through E by the relationship

$$K^2 = EG$$

125

then K = $\sqrt{(2.94 \times 0.674)}$

  = 1.408 kN mm$^{-3/2}$

Given that K = $\alpha F \ell^{\frac{1}{2}}/w$

  $F = \dfrac{Kw}{\alpha \ell^{\frac{1}{2}}}$

Putting in the relevant data, we have

  $F = \dfrac{1.408 \times 60}{0.90 \times 24.7^{\frac{1}{2}}}$ kN = 18.90 kN

Example 9.2

(a) A small-scale laboratory impact tester is shown in figure 9.6. At the start of the test the striker, which weighs 1.2 kg, is at position A. On striking the specimen the pointer on an attached protractor scale, originally at O, indicates 105° from this position. In other words, the striker is now at position B. Calculate the impact energy of the specimen.

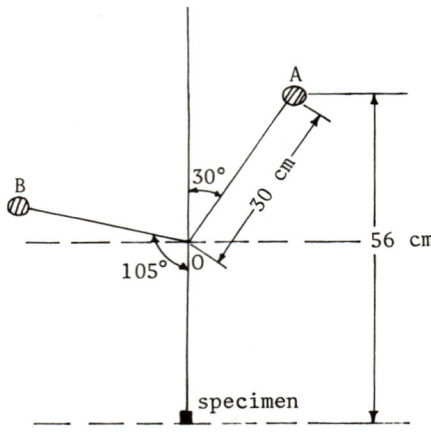

Fig. 9.6

(b) The variation, with notch depth, of the impact energy of wood and Perspex, using this tester is shown in figure 9.7. The simple Griffith equation states that

  impact strength = $\sqrt{(k/\text{notch depth})}$

where k is a constant. On the basis of the above information, explain the difference in impact behaviour between these two materials.

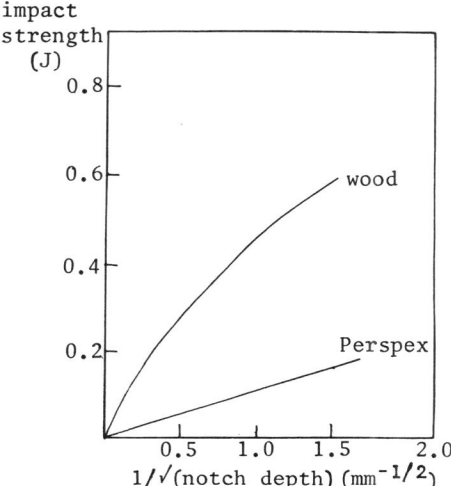

Fig. 9.7

(a) Stored energy before impact = 1.2 × 9.81 × 0.56 = 6.59 J

$$\text{Stored energy after impact} = \frac{(30 + 30 \sin 15°) \times 1.2 \times 9.81}{100}$$

$$= 4.45 \text{ J}$$

Impact energy = 6.59 - 4.45 = 2.14 J

(b) Figure 9.7 shows that Perspex obeys the Griffith equation and is therefore notch-sensitive. Wood does not obey the law and is thus notch-insensitive. The decrease in strength with increasing notch depth is due to the decrease in cross-section.

Example 9.3

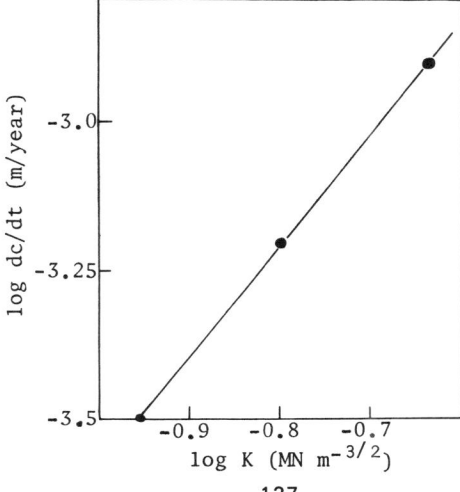

An investigation of the rate of crack growth in cold rolled brass when exposed to a solution of ammonium sulphate and subject to a static stress gave the following results

Nominal stress σ (MNm^{-2})	Crack length c (mm)	Crack growth rate dc/dt (mm/year)
4	0.25	0.3
4	0.50	0.6
8	0.25	1.2

Show that these results are consistent with the crack growth rate being a simple function of the stress intensity factor $K = \sigma\sqrt{(\pi c)}$.

The critical strain energy release rate, $G_{\ell c}$, for unstable fracture of this material in this environment is 55 kJm^{-2} and E = 110 GNm^{-2}. It is proposed to use this brass for piping in an ammonium sulphate plant. The pipes must sustain a tensile hoop stress of 85 MNm^{-2} and experience has shown that longitudinal scratches 0.02 mm deep are likely to occur on the inner surfaces of the pipes. Estimate how long a pipe would last without fracturing once the ammonium sulphate solution started to flow through it. (Cambridge)

Working in MN, m and year units the stress intensity factor, K, is obtained for each nominal stress-crack length combination with the aid of the relation $K = \sigma\sqrt{\pi c}$. The results are

K (MNm$^{-3/2}$)	dc/dt (m/year)
0.1121	0.3×10^{-3}
0.1585	0.6×10^{-3}
0.2242	1.2×10^{-3}

The usual K to dc/dt relationship is of the form

$$dc/dt = BK^n$$

where B and n are constants.

The above results are plotted after the fashion of this relationship (figure 9.8) and yield

$$dc/dt = 0.0239 \, K^2$$

The critical stress intensity factor, $K_{\ell c} = \sqrt{(110 \times 55)}$ MNm$^{-3/2}$

$$= 77.78 \text{ MNm}^{-3/2}$$

Thus the critical crack size $a_c = 0.267$ m, while the initial crack size $a_i = 0.00002$ m. We can thus write

$$\int_{a_i}^{a_c} \frac{dc}{c} = \int_0^T B(\Delta\sigma)^2 \, \pi \, dt$$

where T is the time to fracture.

So

$$T = (\ell n \, 0.267 - \ell n \, 0.00002)/0.0239 \times 85^2 \pi$$

$$= 0.0175 \text{ year} = 6.4 \text{ days}$$

Example 9.4

In a particular structure, two square-section bars must be butt-welded together, as shown in figure 9.9. The weld does not penetrate completely and leaves a central disc-shaped cavity. Only two steels, I and II, are available in the required size. Assuming that plane-strain conditions operate and that the design stress, σ, is 60% of the yield stress σ_y, determine which steel will be the safer in use.

Take the flaw shape parameter, Q, to be 1.00 and refer to figure 9.10.

Steel	σ_y (MN m^{-2})	Fracture Toughness, K_{Ic} (MN m$^{-3/2}$)
I	300	120
II	320	100

	Values of Q					
$c/2\ell$	σ/σ_y					
	<0.6	0.6	0.7	0.8	0.9	1.0
0.1	1.10	1.02	0.98	0.95	0.91	0.88
0.2	1.29	1.22	1.17	1.15	1.12	1.07
0.3	1.60	1.52	1.48	1.45	1.41	1.38
0.4	1.98	1.90	1.87	1.83	1.79	1.76

Assuming that the cracks and the plastic zones associated with them are small compared to the dimensions of the material, we can write

$$\text{fracture stress, } \sigma_f^2 c = K_{Ic}^2 \alpha$$

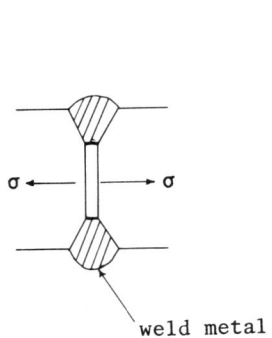

σ ← → σ

weld metal

Fig. 9.9

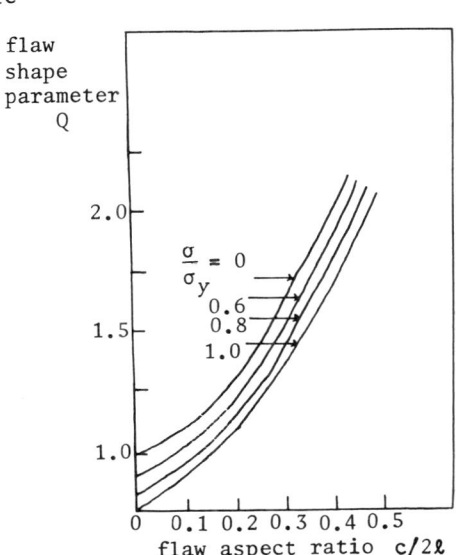

flaw shape parameter Q

$\dfrac{\sigma}{\sigma_y} = 0$

0.6
0.8
1.0

flaw aspect ratio $c/2\ell$

Fig. 9.10

where $\alpha = Q/1.21\pi$ for surface flaws. Q, the flaw shape parameter, is a complex function of σ/σ_y and $c/2\ell$ (figure 9.10). Fracture stress

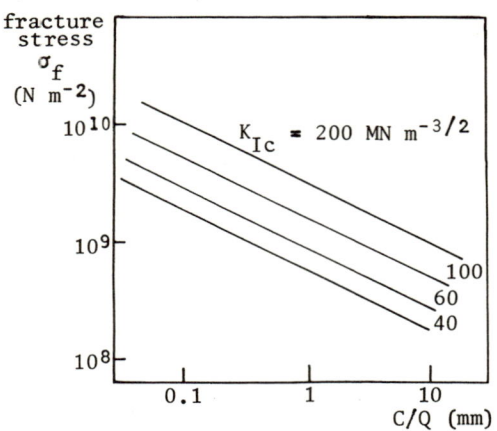

Fig. 9.11

is also a function of c/Q (figure 9.11). Thus we can write

$$\frac{c}{Q} = \frac{1}{1.21\pi} \left(\frac{K_{Ic}}{\sigma}\right)^2$$

With $\sigma/\sigma_y = 0.6$, $Q = 1.00$, $c/2\ell = 0.09$, we can tabulate the results as follows.

Steel	K_{Ic} (MN m$^{-3/2}$)	σ (MN m^{-2})	c/Q (m)	c (m)	2ℓ (m)
I	120	180	0.116	0.116	1.29
II	100	192	0.071	0.071	0.79

It is thus seen that cracks of greater depth and length can be tolerated in steel I at almost comparable working stresses as with steel II, so steel I is safer.

Example 9.5

Choose between the three materials, whose properties are tabulated below, for an application which maximises safety and minimises weight.

Material	Density, ρ (Mg m^{-3})	Yield stress, σ_y (MN m^{-2})	Fracture Toughness, K_{Ic} (MN m$^{-3/2}$)
A	7.86	1730	110
B	2.70	587	33
C	4.51	965	88

Assume that (a) simple inspection methods will detect and remove cracks larger than c = 4 mm and 2ℓ = 20 mm (see figure 9.12) and that therefore cracks of this size are the largest to be found in the material; and (b) the design stress is chosen to be half the yield stress of the material. (Refer to figures 9.10 and 9.11.)

130

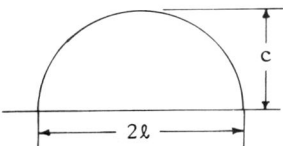

Fig. 9.12

Given

(a) $\dfrac{\sigma, \text{ design stress}}{\sigma_y, \text{ yield stress}} = 0.5$

(b) $c/2\ell = 0.2$

Then from figure 9.10 $Q = 1.29$ and hence $c/Q = 3.1$ mm. The fracture stress, σ_f, for the various material is then obtained from figure 9.11. The results are tabulated below.

Material	Design Stress, σ (MN m$^{-\frac{1}{2}}$)	Fracture Stress, σ_f (MN m^{-2})	Factor of Safety[*]
A	865	1100	1.27
B	295	300	1.02
C	483	800	1.66

[*]Here the factor of safety is defined as σ_f/σ. It is then seen that the largest factor of safety is obtained with material C, which has the second lowest density. This use of material C will maximise safety and minimise weight.

Example 9.6

A design feature which requires careful attention is that of a keyway or spline in a shaft subject to torsion. An approximation to the above is a shallow longitudinal groove with a semi-circular root. Neuber has given the following expression for the shear stress concentration factor, K_{ts}, in this case

$$K_{ts} = \frac{\tau_{max}}{\tau_{nom}} = 1 + \sqrt{\left(\frac{t}{\rho}\right)}$$

where τ_{max} = maximum shear stress, τ_{nom} = nominal shear stress, t = depth, ρ = radius of the groove.

For a groove with a 3 mm radius and 7.5 mm deep, in a shaft 6 cm diameter subjected to a torque of 2 kN m, calculate the maximum shear stress in the shaft due to the discontinuity,

From simple theory of torsion

$$\frac{T}{J} = \frac{G\theta}{l} = \frac{2\tau}{d}$$

131

$$\tau_{nom} = \frac{16T}{\pi d^3} = \frac{16 \times 2 \times 10^3 \times 10^6}{\pi \times 6^3} = 47 \text{ MN m}^{-2}$$

$$K_{ts} = 1 + \sqrt{\left(\frac{t}{\rho}\right)} = 1 + \sqrt{\left(\frac{7.5}{3}\right)} = 2.58$$

Now $K_{ts} = \tau_{max}/\tau_{nom}$, therefore

$$\tau_{max} = K_{ts} \times \tau_{nom}$$

$$= 2.58 \times 47$$

$$= 121.26 \text{ MN m}^2$$

Example 9.7

A 10 cm wide plate of thickness 5 mm has a transverse hole of 35 mm diameter drilled in it. The plate is subjected to a tensile force of 170 N. Using the stress concentration factor data given below, calculate the maximum value of the tensile stress in the section.

Hole diameter / plate width	0.1	0.2	0.3	0.4	0.5
Stress concentration factor	2.72	2.51	2.37	2.25	2.17

Using linear interpolation, the stress concentration factor, K_t, for (hole diameter)/(plate width) of 0.35 is 2.31. Now $K_t = \sigma_{max}/\sigma_{nom}$

and $$\sigma_{nom} = \frac{\text{applied tensile force}}{(\text{plate width} - \text{hole diameter}) \times \text{thickness}}$$

Thus $$\sigma_{nom} = \frac{170 \times 10^6}{65 \times 5} = 0.52 \text{ MN m}^{-2}$$

and $$\sigma_{max} = 2.31 \times 0.52 = 1.20 \text{ MN m}^{-2}$$

Example 9.8

The design criterion for the working stress of a brittle structure is governed by the sensitivity of crack detection. A high-strength steel plate, 3 m × 3 m × 0.01 m containing an edge crack 0.2 m long failed under a uniform tensile load of 2×10^6 N. If cracks greater than 5×10^{-4} m deep can easily be detected, what is the maximum load that the structure will withstand?

The Griffith crack equation is

$$\sigma_f = \sqrt{\left(\frac{2\gamma E}{\pi C}\right)}$$

Here $\dfrac{2 \times 10^6}{9} = \sqrt{\left(\dfrac{2\gamma E}{\pi \times 0.2}\right)}$

thus $2\gamma E = \dfrac{4 \times 10^{12} \times \pi \times 0.2}{81} \, N^2 \, m^{-3}$

Putting this value in the equation for the next case, we have

$$\sigma_f = \sqrt{\left(\dfrac{4 \times 10^{12} \times \pi \times 0.2}{81 \times \pi \times 5 \times 10^{-4}}\right)}$$

$$= 4.47 \, MN \, m^{-2}$$

whence load $= 4.47 \times 9 = 40 \, MN$

Example 9.9

Freshly annealed glass containing surface flaws of length 0.1 μm breaks under a tensile stress of 120 MN m⁻². If a sample of this glass is now subjected to a stress of 30 MN m⁻², failure is found to occur after 10 days under stress. Assuming that the surface energy does not change, calculate the average rate at which the crack has grown during the period of the test.
(Nottingham)

From Griffith crack theory, we have

$$\sigma_f = \sqrt{\left(\dfrac{2\gamma E}{\pi C}\right)}$$

Applying this to the first case, we have

$$120 \times 10^6 = \sqrt{\left(\dfrac{2\gamma E \times 10^9}{\pi \times 0.05 \times 10^{-6}}\right)}$$

whence $2\gamma E = 2262 \times 10^{-3} \, N^2 \, m^{-3}$. Now putting this in the second case, we have

$$30 \times 10^6 = \sqrt{\left(\dfrac{2262 \times 10^6}{\pi C \times 10^{-6}}\right)}$$

whence $C = 0.8$ μm. The crack has grown from 0.1 μm to 1.6 μm in 10 days, so that crack growth rate is 0.15 μm per day.

Example 9.10

The fracture toughness, G_c, of a material may be defined as the force per unit length of crack to propagate the crack rapidly. For infinitely wide plates it is analogous to P in the Orowan equation. For finite-width cracks it may be defined as

$$G_c = \dfrac{\sigma^2 L}{E}(1 - \gamma^2) \, \tan\left(\dfrac{\pi c}{L}\right)$$

133

where 2c = crack length, σ = fracture stress, L = plate, γ = Poisson's ratio.

It is required that a steel plate have a fracture toughness of at least 10 MN m$^{-3/2}$. Investigate whether this requirement is fulfilled for a thin steel plate 30 cm wide containing a central crack 12.5 mm long with a fracture stress of 625 MN m^{-2}. Young's modulus = 200 GN m^{-2}; Poisson's ratio = 0.3.

$$G_c = \frac{\sigma^2 L}{E}(1 - \gamma^2)\ \tan\left(\frac{\pi c}{L}\right)$$

With the values of the various parameters, we have

$$G_c = \frac{625^2 \times 0.3}{200 \times 10^3}(1 - 0.3^2)\ \tan\left(\frac{\pi \times 6.25}{300}\right)$$

$$= 0.61\ \text{kN m}^{-1}$$

For plane stress conditions

$$K_{Ic} = \sqrt{(EG_c)} = \sqrt{(200 \times 10^9 \times 0.61 \times 10^3)}$$

$$= 11\ \text{MN m}^{-3/2}$$

For plane strain conditions

$$K_{Ic} = \sqrt{\left(\frac{EG_c}{(1 - \gamma^2)}\right)} = \sqrt{\left(\frac{200 \times 10^9 \times 0.61 \times 10^3}{0.91}\right)}$$

$$= 11.6\ \text{MN m}^{-3/2}$$

Thus the requirement of a fracture toughness of least 10 MN m$^{-3/2}$ is fulfilled whichever stress conditions apply.

Example 9.11

A sharp penny-shaped crack with a diameter of 2.5 cm is completely embedded in a solid. Catastrophic failure occurs when a stress of 700 MN m^{-2} is applied.

(a) What is the fracture toughness of the material? (Assume that this value is for plane strain conditions.)
(b) If a sheet (0.75 cm thick) of this material is prepared for fracture toughness testing (t = 0.75 cm, a = 3.75 cm), would the facture toughness value be a valid test number? (The yield strength of the material is 1100 MN m^{-2}.)

(a) The fracture toughness, $K_{Ic} = \sigma\sqrt{(\pi a)}$

or $K_{Ic} = 700\ \sqrt{(\pi \times 1.25 \times 10^{-2})}$

$$= 138.73\ \text{MN m}^{-3/2}$$

(b) For valid K_{Ic} result

$$t \text{ and } a \geqslant 2.5 \left(\frac{K_{Ic}}{\sigma_{yield}}\right)^2$$

In this case

$$2.5 \left(\frac{K_{Ic}}{\sigma_{yield}}\right)^2 = 2.5 \left(\frac{138.73}{1100}\right)^2 = 4 \text{ cm}$$

Both t and a are less than 4 cm, so the K_{Ic} result is invalid.

Example 9.12

A compact tension test specimen (a/w = 0.5), is designed and tested according to standard materials testing procedure. Accordingly, a Type 1 load against displacement (P = δ) test record was obtained and a measure of the maximum load P_{max} and a critical load measurement point P_Q were determined. The specimen dimensions were determined as w = 10 cm, t = 5 cm; the critical load point measurement point P_Q = 100 kN and P_{max} = 105 kN. Assuming that all other standard testing procedure requirements (regarding the establishment and sharpness of the fatigue crack starter) were met, determine the critical value of stress intensity.

Does it meet conditions for a valid answer, if the material yield stress is 700 MN m^{-2}?

For a compact tension test specimen, the fracture toughness is given by

$$K_{IQ} = \frac{P}{t\sqrt{(w)}}[29.6\left(\frac{a}{w}\right)^{1/2}-185.5\left(\frac{a}{w}\right)^{3/2}+655.7\left(\frac{a}{w}\right)^{5/2}-1017\left(\frac{a}{w}\right)^{7/2}+638.9\left(\frac{a}{w}\right)^{9/2}]$$

With a/w = 0.5 the term in the brackets = 9.6. From fracture mechanics theory, with P_{max}/P_Q (=105/100) less than 1.1, K_{IQ} is computed with P = P_Q and K_{IQ} is equal to K_{Ic}. With P_Q = 100 kN, t = 5 cm and w = 10 cm, K_{Ic} = 61 MN m$^{-3/2}$.

For valid results

$$t \text{ and } a \geqslant 2.5 \left(\frac{K_{IQ}}{\sigma_{yield}}\right)^2$$

In this case, 2.5 $(K_{IQ}/\sigma_{yield})^2$ = 0.019 m and t and a are both greater than this quantity, so the K_{Ic} answer is valid.

Example 9.13

A large steel plate is used in an engineering structure. A vandal
intent on destroying the component decides to cut a very sharp notch
in the edge of the plate (perpendicular to the loading direction).
If he walks away from the scene at 5 km h^{-1}, how far away can he get
before his plan succeeds? Use the following data.

(a) The plate is cyclically loaded from 0 to 80 kN at a frequency of
    25 cycles per second.
(b) The plate is 20 cm wide and 0.3 cm thick.
(c) The plane strain fracture toughness of the material is 60 MN m$^{-3/2}$
(d) The initial mutilating mark was 1 cm long.
(e) The material has the following crack growth rate-stress intensity
    factor relationship.

$$\frac{da}{dN} = 1.35 \times 10^{-10} (\Delta K)^{2.25}$$

Change in applied stress, $\Delta\sigma = \dfrac{80 \times 10^3}{6 \times 10^{-4}} = 133$ MN m^{-2}

Thus a_c, critical flaw size for fast fracture, can be estimated from

$$K_{Ic} = 1.12\sigma_{max} \sqrt{(\pi a_c)}$$

i.e. $a_c = \dfrac{60^2}{(1.12)^2 (133)^2 \pi} = 0.052$ m

Now $\dfrac{da}{dN} = 1.35 \times 10^{-10} (\Delta k)^{2.25}$

We can write

$$\int_{0.010}^{0.052} \frac{da}{(1.12 \times 133)^{2.25} a^{1.125} \pi^{1.125}} = 1.35 \times 10^{-10} \int_0^N dN$$

whence N = 70 000 cycles. Time taken to cover this number of cycles
at 25 cycles per second = 70 000/25 s = 0.78 h. The student walks
away at a rate of 5 km per hour, so he will be a distance of 3.9 km
away before his plan succeeds.

Example 9.14

After 2 years of service a 10 cm wide panel of an aluminium alloy
was found to contain a 0.5 cm long edge crack oriented normal to the
applied stress. Given the following material properties and service
conditions: fracture toughness of alloy = 35 MN m$^{-3/2}$; the component
was designed to withstand one start-up/shut-down cycle per day for

20 years (assume that there are 250 operating days in a year); the cyclic stress range is 0 to 70 MN m^{-2}; cyclic growth characteristics of the material can be represented by an equation of the form

$$\frac{da}{dN} = 3.3 \times 10^{-9}(\Delta K)^{3.5}$$

where a = crack size (m), N = number of cycles, ΔK = stress intensity range (MN m$^{-3/2}$): (a) comment on the safety of the product of which the panel is an integral part; and (b) state how the safety is affected if the crack was found to emanate from a rivet hole 50 mm in diameter.

(a) First determine a_c, the critical flow size for fast fracture, from the relationship

$$K_{Ic} = 1.12\sigma_{max} \sqrt{(\pi a_c)}$$

thus $a_c = \dfrac{35^2}{(70 \times 1.12)^2 \pi} = 0.063$ m

now $\dfrac{da}{dN} = 3.3 \times 10^{-11}(\Delta K)^{3.5}$

so we can write

$$\int_{0.005}^{0.063} \frac{da}{(70 \times 1.12)^{3.5} a^{1.75} \pi^{1.75}} = 3.3 \times 10^{-10} \int_{0}^{N} dN$$

whence N = 5.8×10^3 cycles, so the component will not fail before the 20 year life.

(b) For a crack emanating from a hole of radius R

$$K_{Ic} \sim 1.12(3\sigma) \sqrt{(\pi a_c)} \quad \text{for } a < c < R$$

thus $a_c = [\dfrac{35}{(3 \times 70 \times 1.12)}]^2 \dfrac{1}{\pi} = 0.007$ m

so that

$$\int_{0.005}^{0.007} \frac{da}{(1.12 \times 210)^{3.5} a^{1.75} \pi^{1.75}} = 3.3 \times 10^{-10} \int_{0}^{N} dN$$

whence N = 3.25×10^2 cycles, so now the component will fail before the intended 20 years life.

Example 9.15

A welded component is required to have a fatigue life of 100 000 cycles. Determine whether the design is adequate, using the following information.

(a) The component is made of martensitic steel, with yield stress of 689 MN m^{-2} and fracture toughness of 165 MN m$^{-3/2}$.
(b) The maximum initial flaw size detectable by the inspection technique is 7.62 mm (edge crack in tension).
(c) The component is to be subjected to a stress which varies from 310 to 172 MN m^{-2}.
(d) For the geometrical configuration of an edge crack in tension, the stress intensity factor, K_I, and crack size, a, relationship is

$$K_I = 1.12\sigma \sqrt{(\pi a)}$$

where σ = applied stress.
(e) The crack growth rate-stress intensity factor relationship is

$$\frac{da}{dN} = 1.37 \times 10^{-10}(\Delta K_I)^{2.25}$$

where N = number of fatigue cycles.
(f) Use both direct integration and numerical integration, with a crack size increment of 12.70 mm.

Use the relationship $K_{Ic} = 1.12\sigma\sqrt{(\pi a_c)}$ with the subscript c referring to critical conditions. Then

$$a_c = \left(\frac{K_{Ic}}{1.12\sqrt{(\pi)}\sigma_{max}}\right)^2 = \left(\frac{165}{1.12\sqrt{(\pi)} \times (310)}\right)^2 = 71.1 \text{ mm}$$

(a) Direct integration.

Given

$$\frac{da}{dN} = 1.37 \times 10^{-10}(\Delta K)^{2.25} \tag{1}$$

We can write

$$\int_{a_i}^{a_c} \frac{da}{(1.12\Delta\sigma)^{2.25}\pi^{1.125}a^{1.125}} = 1.37 \times 10^{-10}\int_{0}^{N} dN$$

a_i = 0.00762 m, a_c = 0.0711 m and $\Delta\sigma$ = 138 MN m^{-2}, thus

$$\int_{0.0076}^{0.071} a^{-1.125} da = 4.19 \times 10^{-5} N$$

138

whence N = 86 000 cycles.

(b) Step-by-step integration. Determine ΔK_I with a_{av} representing the average crack size between the two crack size increments a_i and a_j

$$K_I = 1.12\Delta\sigma \; \sqrt{(\pi a_{av})} = 1.98(138) \sqrt{a_{av}}$$

From equation 1 solve for ΔN for each crack size increment, replacing da/dN by $\Delta a/\Delta N$

$$\Delta N = \frac{\Delta a}{1.37 \times 10^{-10} \, (1.98 \times 138 \, \sqrt{a_{av}})^{2.25}} \quad \text{cycles}$$

For the first step

$$a_{av} = \left(\frac{20.32 + 7.62}{2}\right) = 13.97 \text{ mm}$$

whence ΔN = 37 400 cycles.

The above step is repeated for crack size increment from 20.32 mm to 33.02 mm, etc., and ΔN evaluated at each step. Then $\Sigma\Delta N$ is obtained. The final results are tabulated below.

a_i (mm)	a_f (mm)	a_{av} (mm)	ΔK (MN m$^{-3/2}$)	ΔN (cycles)	$\Sigma\Delta N$ (cycles)
7.62	20.32	13.97	32.24	37 400	37 400
20.32	33.02	26.67	44.54	18.100	55 500
33.01	45.72	39.37	54.10	11 700	67 200
45.72	58.42	52.07	62.30	8 500	75 700
58.42	71.12	64.77	69.40	6 700	82 400

If the required life is 100 000 cycles, then this design is inadequate.

PROBLEMS

(1) For an elliptical hole in a plate the value of the stress concentration factor was shown by Inglis to be (1 + 2a/b), where a and b are the minor and major axes of the ellipse, respectively. For a scratch having an elliptical bottom of radius ρ and depth δ, what is the value of the stress concentration factor?

$$[1 + 2 \sqrt{(\delta/\rho)}]$$

(2) Use the Neuber nomographs shown in figure 9.13 and the information in figure 9.14 to determine the stress concentration factors for a shaft with a groove of zero flank angle, shown in figure 9.14, when subjected to (a) cyclic bending stresses and (b) cyclic torsional stresses.

$$[3; 1.9]$$

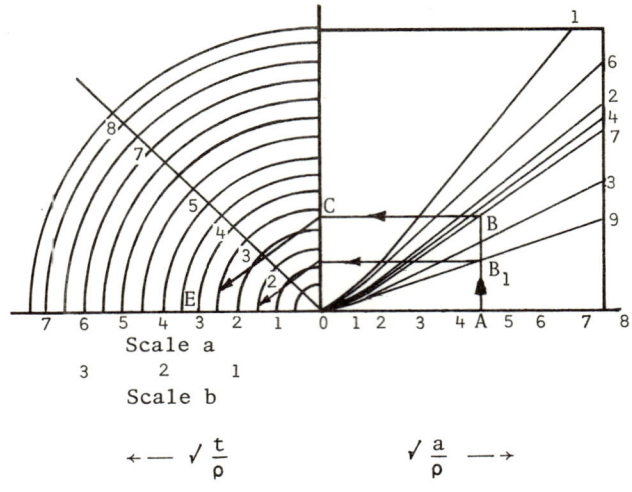

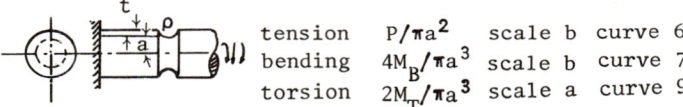

tension $P/\pi a^2$ scale b curve 6
bending $4M_B/\pi a^3$ scale b curve 7
torsion $2M_T/\pi a^3$ scale a curve 9

Fig. 9.13

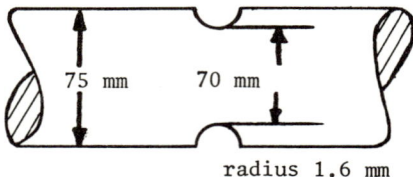

radius 1.6 mm

Figure 9.14

(3) An aircraft wing is to be fabricated from an age-hardened aluminium alloy which has a yield stress, σ_y, of 340 MN m^{-2}. Non-destructive testing (NDT) facilities are available which can detect cracks when they reach a size of 3 mm or more. Calculate whether these facilities are adequate for detecting flaws of allowable size in the wing.

It is suggested that the strength/weight ratio of the wing could greatly be improved by fully heat-treating the alloy to increase σ_y to 480 MN m^{-2}. Investigate whether this heat treatment can be undertaken in the NDT equipment. Assume plane strain conditions. Plane

140

strain fracture toughness of alloy in initial condition and after full
full heat treatment are 45 and 21 MN m$^{-3/2}$. Take design stress
= $1/2\sigma_y$.

If you are now given the additional information that the original
material consists of 10 mm thick sheets investigate whether plane
strain conditions do in fact apply in both cases.
(Nottingham)   [yes; yes; no; no]

(4) A metallic material containing initial flaws no greater than
200 μm long is stressed at 600 MN m^{-2} in a corrosive environment.
Determine the minimum value of K$_{I.S.C.C.}$ needed for this material to
avoid stress corrosion cracking.

$$[10.64 \text{ MN m}^{-3/2}]$$

(5) For a centre-cracked plate specimen the stress intensity factor
is

$$K_I = 1.77 \frac{Pa^{1/2}}{Bw} [1 - 0.1\left(\frac{2a}{w}\right) + \left(\frac{2a}{w}\right)^2] \text{ N mm}^{-3/2}$$

where P = tensile load, B × w = cross-section, 2a = crack length.
For an aluminium alloy with yield stress of 540 N mm^{-2} and
K$_{Ic}$ = 940 N mm$^{-3/2}$, determine (a) the average stress for crack
extension if w = 10B = 20a = 100 mm; and (b) the minimum value of B
and 2a for which the above formula is applicable, using this material.
$$[237.5 \text{ N mm}^{-2}; \text{ B}, 2a \geq 7.57 \text{ mm}]$$

10   MISCELLANEOUS TOPICS

10.1 THE LEVER RULE

The lever rule is a graphical method used in determining the amount
of phases present in a system. The horizontal segment of an isotherm
connecting two phases which are in equilibrium in a two-phase region
is known as a 'tie-line' (figure 10.1). This tie-line, according to
the rule, may be considered to be a lever whose fulcrum is placed at
the total composition and whose arms carry weights (i.e. relative
amounts of phases) inversely proportional to their lengths.

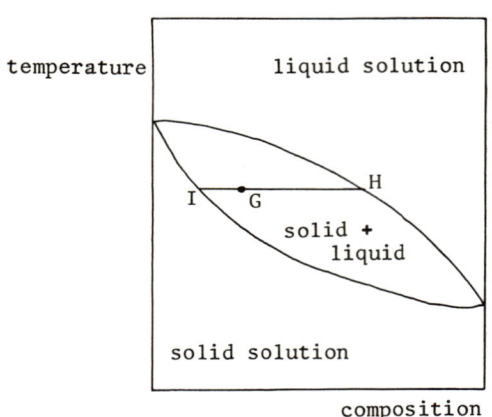

Figure 10.1

10.2 THE TEMPERING PARAMETER

Using principles similar to those applied in creep stress rupture
predictions, tempering response can be estimated. The parameter which
governs the relationship between tempering time and temperature is

$$T(C + \log t)$$

where T = absolute tempering temperature, C is a constant
(15 < C < 20) with a generally used value of 18, t = tempering time
(h). With tempered hardness of the material being obtained after
each tempering operation, a master tempering curve can be obtained as
a plot of hardness versus the tempering parameter (figure 10.2). From
this tempering temperature specification needed to produce a required
hardness after a known tempering time can be obtained.

   Response to tempering is significantly affected by the composition
of the steel, the size of the pieces and the presence of high-temper-

142

ature transformation products and retained austenite. Thus if master tempering curves are constructed from the published literature, such curves would be rigorously applicable only to testpieces of identical composition, size and heat treatment. To overcome such limitations, it is necessary to make corrections to the master curves, using data from some sample testpieces under actual laboratory tempering conditions.

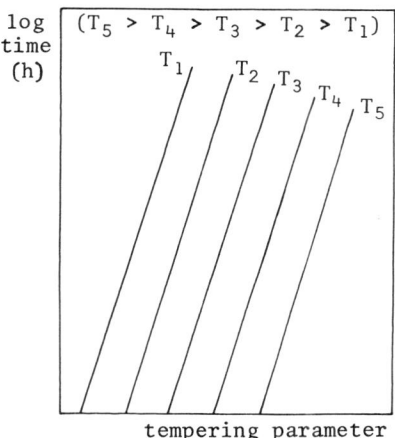

Figure 10.2

10.3 THERMALLY ACTIVATED PROCESSES

There exist many physical situations in which an atom or particle can move from a metastable state to another of greater stability by passing through an intermediate state of higher energy (figure 10.3).

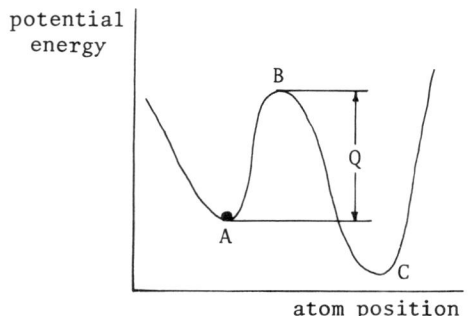

Figure 10.3

The atom at position A can reach a more stable position at C only by passing through the unstable position B with the aid of the necessary additional energy. This additional energy, Q, may take the form of heat, light, etc.

The rate at which such a change can occur in a given atom population is a function of the magnitude of Q and the number of atoms in that population that possess energy at least as great as Q. The number of such atoms is proportional to exp (-Q/RT). Thus

$$\text{reaction rate} = A \exp \left(-\frac{Q}{RT}\right)$$

This equation is known as the Arrhenius rate law. There are a number of processes which are thermally activated and for which, therefore, the Arrhenius relationship is valid. These include diffusion, thermionic emission and corrosion.

Example 10.1

Several specimens of mild steel are cold-worked to the same extent. One is immediately placed in boiling water and it shows a higher yield point after 15 min. After what time would the same recovery of yield point be found in a specimen kept at 15 °C after cold-working? For how much longer can the recovery of the yield point be postponed by placing a specimen in a refrigerator at 0 °C? The activation energy for the diffusion of carbon in α-iron is approximately 7.5×10^7 J kmol^{-1}.
(Cambridge)

(a) Recovery of yield point is regarded as a thermally activated process. Then

$$\text{rate} = A \exp \left(-\frac{Q}{RT}\right)$$

Let rate = k/t (where k is a constant and t = time in minutes). Then considering cases 1 (t = 15 min, T = 373 K) and 2 (t = t min, T = 288 K), we can write

$$\frac{k}{15} = A \exp \left(-\frac{7.5 \times 10^4}{8.314 \times 373}\right) \tag{1}$$

$$\frac{k}{t} = A \exp \left(-\frac{7.5 \times 10^4}{8.314 \times 288}\right) \tag{2}$$

From equations 1 and 2, we have

$$\frac{t}{15} = \exp (7.14) = 1261.4$$

whence t = 13 days.

(b) Considering cases 1 (t = 15 min and T = 373 K) and 3 (t = t´ min and T = 273 K), we can write

$$\frac{t´}{15} = \frac{\exp (-24.18)}{\exp (-33.04)} = 7044$$

whence t´ = 73 days. Thus additional time = 60 days.

Example 10.2

For a precipitation-hardening alloy in the 'as-quenched' condition, the first appearance of a precipitate in the super-saturated solid solution can be detected 1 h after quenching when the temperature of the alloy is held at 15 °C. If the temperature after quenching is raised to 100 °C the precipitates can be detected after 1 min. If it is desired to retard the precipitation hardening process so that no precipitates can be detected in under 1 day, to what temperature must the alloy be cooled after quenching?
(Cambridge)

The appearance of a precipitate in a super-saturated solid solution is regarded as a thermally activated process. Thus we can write

$$\text{rate} = A \exp\left(-\frac{Q}{RT}\right)$$

$$\log k - \log t = \log A - \frac{Q}{2.3 \times RT} \tag{1}$$

where k is a constant, t = time in minutes. Applying this to the first two cases, we have

$$\log k - \log 60 = \log A - \frac{Q}{2.3 \times 8.314 \times 288}$$

$$\log k - \log 1 = \log A - \frac{Q}{2.3 \times 8.314 \times 373}$$

whence $Q = 4.24 \times 10^4$ J mole^{-1}.

Applying equation 1 to cases 1 (t = 1 min) and 3 (t = 60 × 24 min), we have

$$\log k - \log 1 = \log A - \frac{42400}{2.3 \times 8.314 \times 373}$$

$$\log k - \log (60 \times 24) = \log A - \frac{42400}{2.3 \times 8.314T}$$

whence T = 243 K or -30 °C.

Example 10.3

A common mode of failure in many ceramics is known as static fatigue in which the time to failure of the materials at a given stress is dependent on the thermally activated diffusion of cations through the lattice. In soda-lime glass such as is used for making bottles, the activation energy for this process is 65 kJ mol^{-1}. It is found that, at the pressure at which soft drinks bottles are filled, the most likely time to failure is 6 months at 290 K. Estimate by how much this time to failure changes if the bottles are stored in a poorly

ventilated warehouse in which the summer temperature is liable to rise to 310 K. (Molar gas constant = 8.314 J mole^{-1} K^{-1}.)
(Nottingham)

The Arrhenius equation for the rate of thermally activated processes is

$$\text{rate} = A \exp\left(-\frac{Q}{RT}\right)$$

Now rate is inversely proportional to time so we can write for the two cases (at T = 290 K and 310 K)

$$\frac{\text{rate (290 K)}}{\text{rate (310 K)}} = \frac{\exp\left[-65\ 000/(8.314 \times 290)\right]}{\exp\left[-65\ 000/(8.314 \times 310)\right]} = 0.176$$

Thus $\dfrac{\text{time (310 K)}}{\text{time (290 K)}} = 0.176$

$$\text{time (310 K)} = 0.176 \times 6 \text{ months} \simeq 1 \text{ month}$$

whence time change = 5 months.

Example 10.4

Some values quoted for the diffusion coefficient, D, of carbon into γ-iron, determined from tests on almost pure iron with a gaseous carburising agent, are as follows.

Temperature (°C)	D (m^2 s^{-1} × 10^{12})
800	1.5
900	7.5
950	11.8
1000	20.0
1050	28.0
1100	45.0

Examine whether these values follow an Arrhenius rate law relationship. Suggest a reason for any that do not.

Estimate a value for the activation energy for the diffusion of carbon in γ-iron.
(Cambridge)

Assuming diffusivity follows an Arrhenius rate relationship, then we can write

$$D = A \exp\left(-\frac{Q}{RT}\right)$$

So a plot of log D against 1/T would give slope (= -Q/2.3R) and intercept (=A). From figure 10.4 we have slope = -8000 K whence Q = 153 kJ mole^{-1}.

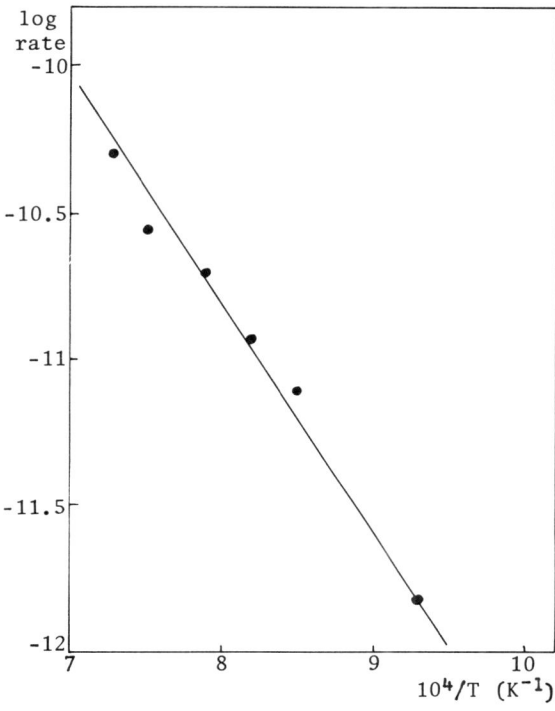

log
rate

Fig. 10.4

Example 10.5

Hydrogen iodide is found to decompose into hydrogen and iodine at elevated temperatures according to the reaction

$$2HI = H_2 + I_2$$

The rate of gas production for a given mass of hydrogen iodide decomposing is found experimentally to vary with temperature as follows.

Temperature (°C)	Rate of Gas Production $(10^{-10} \ m^3 \ mole^{-1} \ s^{-1})$
283	3.52
302	12.20
356	302.00
393	2190.00
427	11600.00
508	395000.00

Estimate the activation energy of the reaction. (R, molar gas constant = 8.314 J mole^{-1} K^{-1}.)

We make the assumption that the rate of gas production in the decomposition of hydrogen iodide into hydrogen and iodine is a thermally activated process. Thus we can write

$$\text{rate} = A \exp\left(-\frac{Q}{RT}\right)$$

where A is a constant, Q = activation energy, R and T have their usual meanings. A plot of log rate against $1/T$ then gives the slope = $-Q/2.3R$, from which Q can be obtained. The results are now put in a suitable form for plotting.

log rate	-9.45	-8.91	-7.52	-6.66	-5.94	-4.40
$10^3/T$ (K^{-1})	1.80	1.74	1.59	1.50	1.43	1.28

The plot (figure 10.5) has a slope of -9711 K, whence $Q = 2.3 \times 8.314 \times 9711 = 186$ kJ mole^{-1}.

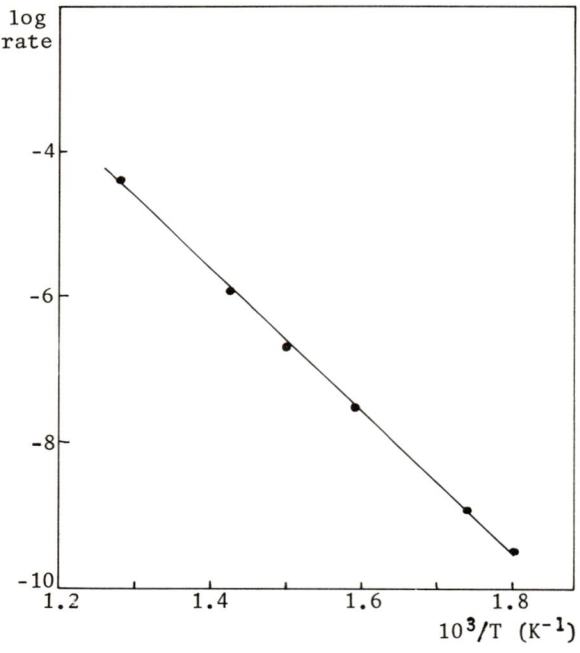

Fig. 10.5

Example 10.6

In a study of slowly propagating cracks in high-strength steel plates under constant stress, it was found that in a moist air environment the crack growth rate increased with temperature as follows.

Growth rate (m s^{-1} × 10^5)	0.70	2.20	8.70	29.1
Temperature (K)	278	298	328	360

148

Show that for these conditions crack propagation is a thermally activated process and hence determine the activation energy.

From an inspection of the diffusion data given below, outline the probable cause of the embrittlement of the steel.

Diffusing Element in α-iron	Activation Energy for Diffusion (MJ kmole^{-1})
Hydrogen	38
Nitrogen	72
Carbon	84
Iron	285

(Cambridge)

The Arrhenius rate equation is

$$\text{rate} = A \exp\left(-\frac{Q}{RT}\right)$$

Thus from a graph of log rate against $1/T$ we can obtain Q, the activation energy, from the fact that slope of the graph = $-Q/2.3R$. The graph in figure 10.6 gives a slope of -2012.5 K, whence $Q = 38.5$ kJ mole^{-1}.

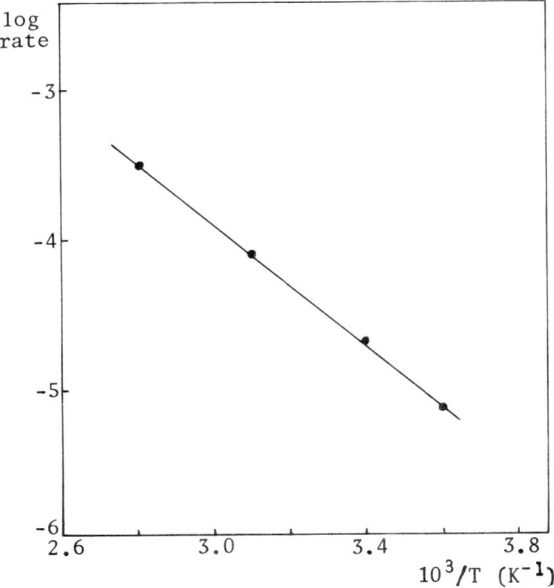

Fig. 10.6

It is obvious that the diffusing element in this case is hydrogen and the probable cause of embrittlement is hydrogen embrittlement.

149

Example 10.7

The time in hours, t, for the recrystallisation of a cold-rolled aluminium strut is given by

$$t = \frac{1}{A \exp\left(-\frac{Q}{RT}\right)}$$

where A is a constant $= 5 \times 10^7$ h, Q = activation energy $= 8.4 \times 10^4$ J mole^{-1}, R = molar gas constant $= 8.314$ J mole^{-1} K^{-1}, T = absolute temperature. If the temperature of the strut in service was constant at 27 $^\circ$C, should concern be shown about loss of strength due to recrystallisation after 20 years' service?

$$t = \frac{1}{A \exp(-Q/RT)}$$

Putting in the values of the various parameters we have

$$\frac{1}{t} = 5 \times 10^7 \exp\left(-\frac{8.4 \times 10^4}{8.314 \times 300}\right) h^{-1}$$

whence t = 0.0847×10^8 h = 967 years. Thus no concern should be shown regarding loss of strength due to recrystallisation in 20 years' service.

Example 10.8

A heavily rolled brass sheet must be annealed for 2 min at 400 $^\circ$C before recrystallisation is 50% complete. How long must the sheet be annealed at 300 $^\circ$C? The activation energy for recrystallisation is 16.8×10^4 J mole^{-1}.

The rate of recrystallisation follows an Arrhenius law. Thus

$$rate = A \exp\left(-\frac{Q}{RT}\right)$$

So $\quad \dfrac{rate\ (673\ K)}{rate\ (573\ K)} = \dfrac{\exp\left(\left[-16.8 \times 10^4/(8.314 \times 673)\right]\right)}{\exp\left(\left[-16.8 \times 10^4/(8.314 \times 573)\right]\right)} = 189$

Now rate of recrystallisation is inversely proportional to time of recrystallisation, thus

$$\frac{time\ (673\ K)}{time\ (573\ K)} = \frac{1}{189}$$

$$time\ (573\ K) = 189 \times 2 = 378\ min$$

Example 10.9

The following results were obtained on a steel tempered for 2 h at the given temperatures.

150

Temperature (°C)	200	315	430	540	630
Rockwell C hardness, R_c (kgf mm^{-2})	54.2	48.3	44.3	36.0	28.0

From this set of data estimate the required temperature to achieve a tempered Rockwell C hardness of 50 if a ¾ h tempering cycle is to be used.

The tempering parameter is T (18 + log t) where T = absolute temperature, t = time (h). This parameter was calculated for each data point and the values are as follows.

Temperature (°C)	Tempering Parameter × 10^{-4}
200	0.87
315	1.08
430	1.29
540	1.49
630	1.65

A graph of hardness against tempering parameter is now plotted (figure 10.7) from which it is found that for a desired hardness of 50 the tempering parameter has a value of 10^4. So that we have

$$T [18 + \log (0.75)] = 10^4$$

whence T = 559 K or 286 °C.

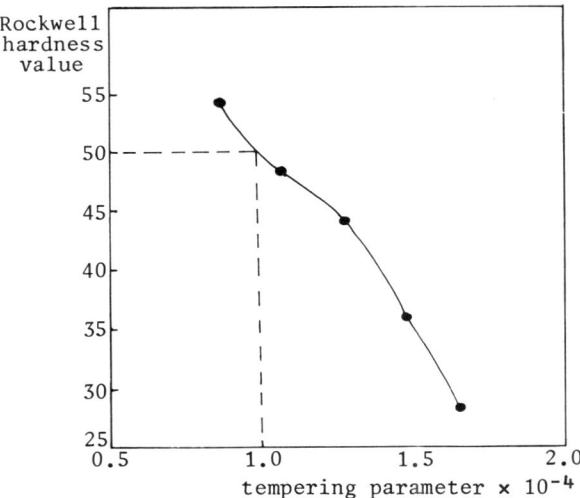

Fig. 10.7

Example 10.10

Use the Lever rule to estimate the percentage liquid and solid phases present in the equilibrium composition (point X in figure 10.8 at the temperature T.)

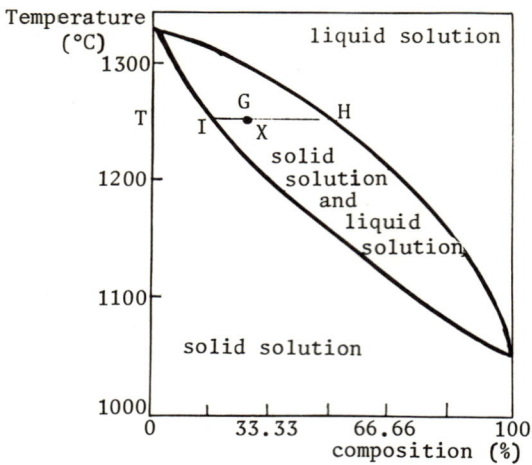

Figure 10.8

By the Lever rule

$\%$ solid phase $= \dfrac{GH}{IH} \times 100 = \dfrac{50 - 27}{50 - 17} \times 100 = 70\%$

$\%$ liquid phase $= \dfrac{IG}{IH} \times 100 = \dfrac{27 - 17}{50 - 17} \times 100 = 30\%$

Example 10.11

A complex shape is to be manufactured from EN 8 steel. Following

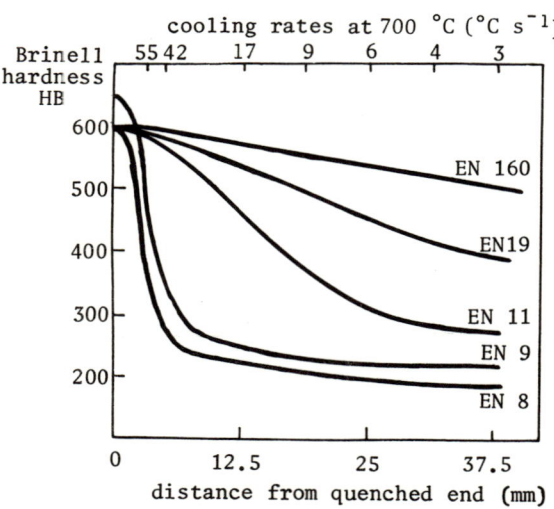

Figure 10.9

152

austenising and quenching in mildly agitated oil, the Brinell hardness 3 mm below the surface is 262HB. This condition is too soft for the required application for which a Brinell hardness of 430HB at 3 mm below the surface is necessary. What steel should be recommended if the austenising and quenching procedures are to remain the same? (Refer to figure 10.9.)

From the Jominy curves for EN 8 steel (figure 10.9), 262HB appears at 6 mm from the water-quenched end. Thus, material at this position cools at the same rate (42 °C s^{-1}) as the material from which the complex shape is made when quenched in agitated oil at 3 mm below the surface.

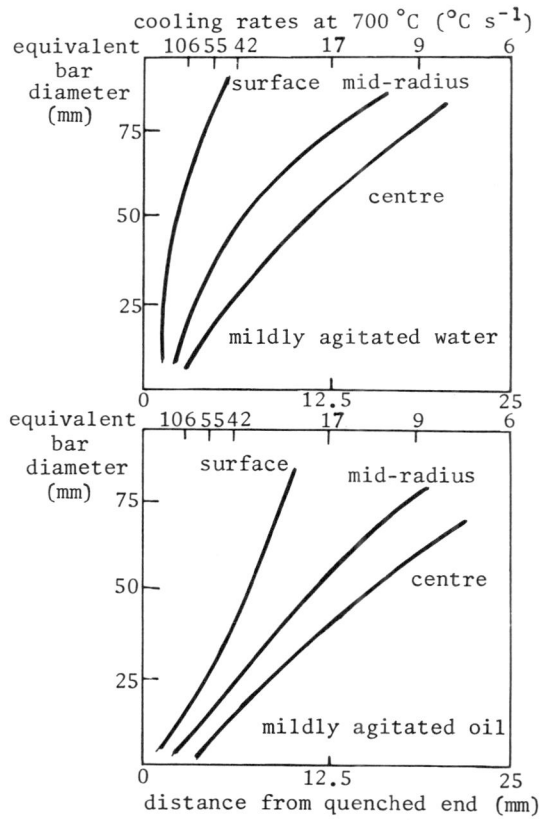

Figure 10.10

Thus we note the hardness obtained for the other steels at the same cooling rate (6 mm station or 42 °C s^{-1}) and observe that EN 160, EN 19 and EN 111 steels all give hardness values greater than 430HB under these quenching conditions.

Example 10.12

How is a Brinell hardness value of 550 obtained 12.5 mm below the

153

surface of a 50 mm diameter bar quenched in (a) mildly agitated water and (b) mildly agitated oil? (Refer to figures 10.9 and 10.10.)

Read from figure 10.10 that the cooling rate at this position (mid radius) is 17 °C s^{-1} and that this corresponds to a Jominy position of 12.5 mm from the quenched end of the bar (figure 10.9). A vertical line is then drawn at this position on figure 10.9 from which it is seen that only EN 160 steel would provide satisfactory hardness, i.e. above 550HB (EN 19 would give almost 550HB).

Example 10.13

From the following information construct and label the equilibrium diagram for the lead-tin system: melting point of lead 327 °C, melting point of tin 232 °C; eutectic temperature 183 °C; eutectic composition 61.9 wt% tin.

Maximum solubility	At eutectic temperature	At room temperature
Tin in lead (α-phase)	19.2 wt% tin	1.3 wt% tin
Lead in tin (β-phase)	2.5 wt% tin	0.1 wt% tin

(Assume straight-line relationships where necessary.)

The constructed equilibrium diagram is shown in figure 10.11.

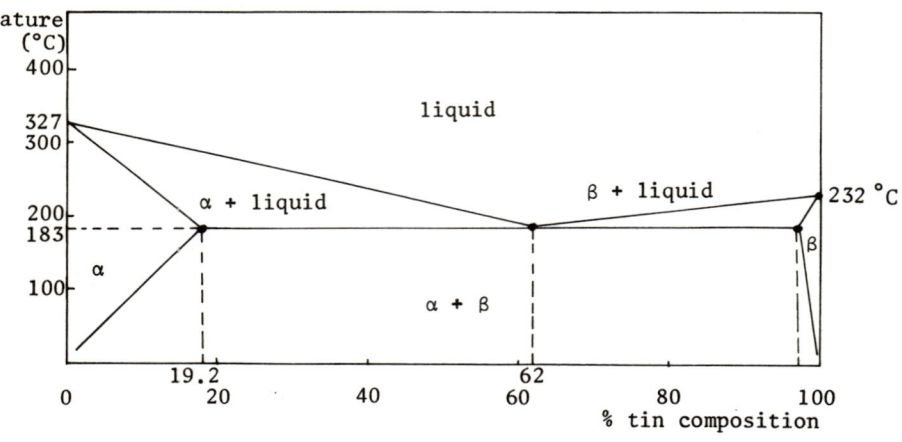

Fig. 10.11

Example 10.14

In an experiment to determine internal stress in electrodeposits using the bent strip method, a steel strip 1 cm × 10 cm × 0.02 cm thick was lacquered on one side and then made the cathode in an electroplating bath until a deposit 0.001 cm thick was produced. When placed on a flat surface, the strip was found to have bent with a deflection at its central point of 0.25 cm. Starting from the following expression, determine the stress in the deposit.

154

$$S = \frac{E}{tR} \int_0^d \left(\frac{2d}{3} - x\right) dx$$

where S = stress, E = Young's modulus of both the substrate and deposit metal (120 GN m^{-2}), t = deposit thickness strip, d = thickness R = radius of curvature of the bent strip.

The expression for the stress, S, is

$$S = \frac{E}{tR} \int_0^d \left(\frac{2d}{3} - x\right) dx$$

which on integration gives

$$S = \frac{Ed^2}{6tR}$$

Now E = 120 GN m^{-2} = 12 $\times$ 10^6 N cm^{-2}; d = 0.02 cm; t = 0.001 cm; R = 0.25 cm. Thus stress in the deposit is

$$S = \frac{12 \times 10^6 \times (0.02)^2}{6 \times 0.001 \times 0.25} = 3.2 \text{ MN m}^{-2}$$

Example 10.15

A vessel for a chemical plant can be manufactured from either 18Cr-8Ni-3Mo stainless steel or from rubber-lined mild steel. The plant is required to last for 20 years, and whereas the stainless steel would not require any maintenance, the rubber lining of the mild steel vessel would have to be replaced after the 4th, 8th, 12th and 16th year at a cost of £2 000 for each lining. The initial cost of the stainless steel and rubber-lined mild steel vessels are £10 000 and £4 000, respectively and the rate of interest is 6%. Calculate the discounted costs, and discuss the various factors which should be taken into account in arriving at a final decision on the choice of material for constructing the vessel.

$$\text{Discounted cost,} = £C \left(1 + \frac{I}{100}\right)^y$$

where C = initial cost, I = interest rate, y = life of the component.

Stainless steel discounted cost = £10 000$(1 + 0.06)^{20}$ = £32 071

Rubber-lined mild steel discounted cost = £$[4000(1 + 0.06)^{20}$

$+ 2000(1 + 0.06)^{16} + 2000(1 + 0.06)^{12} + 2000(1 + 0.06)^8$

$+ 2000(1 + 0.06)^4] = £27 648$

Factors to consider in making final choice include: (a) cost of shut-
down time, (b) inflation, (c) cost of maintenance, (d) availability
of material in required form and shape. Mild steel is readily
available in the right form; it is cheap and strong but needs protective treatment or maintenance. Stainless steel is relatively
expensive but needs little or no maintenance.

PROBLEMS

(1) It is found that a force will inject a given weight of a
thermosetting polymer into a mould in 30 s at 180 °C and in 80 s
at 160 °C.   If the viscosity of the polymer follows an Arrhenius
law, with a rate of process being proportional to exp $(-Q/RT)$, where
Q is the activation energy and R and T have their usual meanings,
calculate how long the process will take at 230 °C.

[3.54 s]

(2) A severely cold-worked metal was found to be 50% recrystallised
at the following combinations of heating times and temperatures:
1 min, 162 °C; 100 min, 97 °C. Consider that the recrystallisation
is a self-diffusion process and estimate the temperature required for
50% recrystallisation in 1 week.

[49 °C]

(3) The instructions for a popular brand of glue recommend that
maximum strength joints can be achieved in 30 min at 422 K, 3 h at
352 K and 3 days at 298 K. Estimate the activation energy responsible
for the cross-linking process responsible for the 'setting' of this
glue. How long will the strongest joint take to form at body temperature?

[41 kJ mole^{-1}; 30 h]

(4) Applying the principle of discounted cash flow and assuming an
interest rate of 8% on invested money, calculate which of these two
schemes of hot dip galvanising is more economical.

    Scheme (a): The steel is grit-blasted to increase the thickness of
the galvanising layer, which is 0.2 mm thick, and will thus require
no maintenance during the contemplated life of 24 years for the
structure. Initial cost of galvanising is £29.50 per tonne.

    Scheme (b): The steel is pickled and galvanised to give a coating
of only 0.10 mm thickness at a cost of £22 per tonne but has to be
painted after 12 years at a cost of £17.50 per tonne.
                    [Scheme A cost, £187.06; Scheme B cost, £183.58]

(5) A cargo aircraft is being designed, and part of the design
requires 80 bolts which are 75 mm long and can withstand a load of
450 N. The bolts can be loaded to 90% of their elastic limit. Two
materials are being considered, aluminium 14S-T6 alloy and annealed
SAE 51410 steel. The aluminium alloy costs £6.60 per kg while the
steel costs £2.20 per kg. The cost of the cargo that the plane can

carry will be reduced by an amount equal to the weight of the bolts. Cargo capacity is worth £110 per kg. Comment on the relative merits and demerits of these materials for this application, from the standpoints of weight and space. The following data apply

	Elastic Limit $(MN\ m^{-2})$	Density $(kg\ m^{-3})$
14S-T6	483	2700
SAE 51410 (annealed)	760	7800

[The aluminium bolts are cheaper because of the low density but the steel bolts are better from the space viewpoint because of lower cross-sectional area]

APPENDIX I   Further Reading

CHAPTER 1

American Society for Metals, *The Science of Hardness Testing and its Research Applications* (ASM, 1973)
Cottrell, A.H., *The Mechanical Properties of Matter* (Wiley, 1964)
McLean, D., *Mechanical Properties of Metals* (Wiley, 1962)
Sharp, H.J., (ed.) *Engineering Materials* (Heywood, 1966)
Varga, O.H., *Stress-Strain Behaviour of Elastic Materials* (Interscience, 1966)
Wyatt, O.H., and Dew-Hughes, D., *Metals, Ceramics and Polymers* (Cambridge University Press, 1974)

CHAPTER 2

Brophy, J.H., Rose, R.M., and Wulff, J., *The Structure and Properties of Materials vol II Thermodynamics of Structure* (Wiley, 1964)
Ramsey, J.A., *A Guide to Thermodynamics* (Chapman & Hall, 1971)

CHAPTER 3

Hall, C., *Polymer Materials: an introduction for technologists and scientists* (Macmillan Press, 1981)
Neilsen, L.E., *Mechanical Properties of Polymers* (Van Nostrand Reinhold, 1962)
Schultz, J., *Polymer Materials Science* (Prentice-Hall, 1974)
Ward, I.M., *Mechanical Properties of Solid Polymers* (Wiley, 1971)

CHAPTER 4

John, V.B., *Introduction to Engineering Materials* (Macmillan Press, 1972)
Wyatt, O.H., and Dew-Hughes, D., *Metals, Ceramics and Polymers* (Cambridge University Press, 1974)

CHAPTER 5

Evans, U.R., *Corrosion and Oxidation of Metals* (Edward Arnold, 1960)
Pludek, V.R., *Design and Corrosion Control* (Macmillan Press, 1977)
Scully, J.C., *Fundamentals of Corrosion,* 2nd ed. (Pergamon, 1975)
Shreir, L.L., (ed.) *Corrosion,* vols I and II (Newnes-Butterworth, 1976)
Tomashov, N.D., *Theory of Corrosion and Protection of Metals* (Macmillan, 1966)
West, J.M., *Electrodeposition and Corrosion Processes* (Van Nostrand, 1971)

CHAPTER 6

Kubaschewski, O., and Hopkins, B.E., *Oxidation of Metals and Alloys*, 2nd ed. (Butterworth, 1962)
Pludek, V.R., *Design and Corrosion Control* (Macmillan Press, 1977)
Tomashov, N.D., *Theory of Corrosion and Protection of Metals* (Macmillan, 1966)

CHAPTER 7

Dorn, J.E., (ed.) *Mechanical Behaviour of Materials at Elevated Temperatures* (McGraw-Hill, 1961)
Garofalo, F., and Bain, E.C., *Fundamentals of Creep and Creep-Rupture in Metals* (Collier-Macmillan, 1965)
McClintock, F.A., and Argan, A.S., (eds.) *Mechanical Behaviour of Materials* (Addison-Wesley, 1966)

CHAPTER 8

Duggan, T.V., and Byrne, J., *Fatigue as a Design Criterion* (Macmillan Press, 1977)
Forsyth, P.J.E., *The Physical Basis of Metal Fatigue* (Blackie, 1969)
Harris, W.J., *Metallic Fatigue* (Pergamon, 1961)

CHAPTER 9

Duggan, T.V., and Byrne, J., *Fatigue as a Design Criterion* (Macmillan Press, 1977)
Knott, J.F., *Fundamentals of Fracture Mechanics* (Butterworth, 1973)
Lawn, B.R., and Wilshaw, T.R., *Fracture of Brittle Solids* (Cambridge University Press, 1975)
Liebowitz, H., (ed.) *Fracture - An Advanced Treatise*, 7 vols (Academic Press, 1969-72)

CHAPTER 10

Cahn, R.W. (ed.) *Physical Metallurgy* (North Holland, 1970)
Cottrell, A.H., *An Introduction to Metallurgy* (Edward Arnold, 1967)
Keyser, C.A., *Basic Engineering Metallurgy* (Prentice-Hall, 1959)

APPENDIX II   Table of Error Function

y	Erf y
0.0	0.0000
0.1	0.1125
0.2	0.2227
0.3	0.3286
0.4	0.4284
0.5	0.5205
0.6	0.6039
0.7	0.6778
0.8	0.7421
0.9	0.7969
1.0	0.8427
1.1	0.8802
1.2	0.9103
1.3	0.9340
1.4	0.9523
1.5	0.9661
1.6	0.9764
1.7	0.9838
1.8	0.9891
1.9	0.9928
2.0	0.9953

APPENDIX III   Theoretical Stress Concentration Factors

Type of deformation, and causes of stress concentrations

k_t

I   Flexure and torsion:
  (a) Semi-circular recess on shaft, with ratio of radius
       of recess to diameter of shaft equal to

0.1	2.00
0.5	1.60
1.0	1.20
2.0	1.10

  (b) Fillet, with ratio of fillet radius to diameter
       of equal to

0.0625	1.75
0.125	1.50
0.25	1.20
0.50	1.10

  (c) Transition at a right angle   2.00

  (d) Sharp V-shaped recess   3.00

  (e) Screw thread (metric sizes)   2.50

  (f) Holes, with ratio of hole diameter to transverse
       dimension of the section between 0.1 and 0.333   2.00

  (g) Machining marks on the surface of a component   1.20 - 1.40

II   Torsion:
  (a) Fillet, with ratio of fillet radius to least diameter
       of shaft equal to

0.02	1.80
0.10	1.20
0.20	1.10

  (b) Keyways   1.60 - 2.00

[From N.M. Belyayev, *Strength of Materials,* 12th ed. (Pergamon, 1959]